Law and Economics of International Telecommunications

Under the Auspices of the
Max Planck Institute for Foreign and International Private Law
edited by Prof. Ernst-Joachim Mestmäcker

Volume 2

Prof. Marcellus S. Snow

The International Telecommunications Satellite Organization (INTELSAT)

Economic and Institutional Challenges Facing an International Organization

Nomos Verlagsgesellschaft
Baden-Baden

CIP-Kurztitelaufnahme der Deutschen Bibliothek

Snow, Marcellus S.:
The International Telecommunications Satellite Organization (INTELSAT): Econom. and Institutional Challenges Facing an Internat. Organization / Marcellus S. Snow. – 1. Aufl. – Baden-Baden: Nomos Verl.-Ges., 1987.
(Law and Economics of International Telecommunications; Vol. 2)
ISBN 3-7890-1434-6
NE: GT

1. Auflage 1987

For my brothers

Stephen and Philip

Editor's Preface

The Institute's project on »The Law and Economics of International Telecommunications« includes the role of international organizations in this area. At the request of the Institute, Professor Marcellus S. Snow kindly agreed to analyse the status and present role of INTELSAT in the context of national and international changes of communications technologies and policies. The unique structure and task of INTELSAT are crucial for an understanding of the present and the future of international telecommunications systems.

In the appendix we publish a United States White Paper on New International Satellite Systems. This study, prepared by a Senior Interagency Group, summarizes United States deregulation policies for International Satellite Systems. President Ronald Reagan, in Presidential Determination No. 85–2 of November 28, 1984, determined »that separate international communications satellite systems are required in the national interest.« This finding under the Communications Satellite Act of 1962 paves the way for the authorization of such separate systems by the Federal Communications Commission. Before final authorization, the United States as a party to INTELSAT is under an obligation to coordinate its separate systems under Article XIV of the Intergovernmental Agreement. Professor Snow's analysis of INTELSAT coordination (at pp. 83 et seq.) and the White Paper are to illustrate the interaction of public international law with national telecommunications laws and policies.

Marcellus S. Snow is Professor of Economics at the University of Hawaii at Manoa, with which he has been affiliated since 1974. He is also a Researcher at that institution's Social Science Research Institute. He holds a Ph.D. in economics from the University of California, Berkley; an M.A. from the School of Advanced International Studies, the Johns Hopkins University; an M.S. from the Massachusetts Institute of Technology; and a B.A. from the University of Utah.

His primary research interest is in the regulation and structure of the telecommunications industries of industrialized countries and of international telecommunications. Particular areas of emphasis include satellite systems and the role of telecommunications in economic development.

His previously published books include *International Commercial Satellite Communications* (1976); *Economic and Policy Problems in Satellite Communications* (1977, with Joseph N. Pelton); *Marketplace for Telecommunications: Regulation and Deregulation in Industrialized Democracies* (1986); and *Telecommunications Economics and International Regulatory Policy: An Annotated Bibliography* (1986, with Meheroo Jussawalla).

We recognize gratefully a grant from the Eduard v. Schwartzkoppen-Stiftung, which made this study possible. We also extend to Paul Aliferis, J.D., for his editorial assistance.

Preface

My first in-depth investigation of INTELSAT appeared in 1976. It had emerged from my earlier activities as a financial and political analyst for the Communications Satellite Corporation during 1969–71 and my graduate work in economics at the University of California, Berkeley during 1971–74. At the time, one spoke of INTELSAT's challenges as »growing pains« peculiar to the »adolescence« of a new and unique international organization that had just completed its first decade.

I hope it does not overtax this biological metaphor to suggest that one can now appropriately speak of something like INTELSAT's mid-life crisis. Having established itself as a viable entity, it must adapt to new economic and technological forces that determine the changing environment within which it operates. It is no longer a brash newcomer on the international scene, flush with triumphs that came almost too easily. Its own successes have attracted more recent and perhaps more commercially minded competitors, such as private international satellite and fiber-optic cable firms, with which it must now find an accommodation.

Because of this earlier work and my continuing interest in the regulatory and institutional aspects of international telecommunications, I was delighted to accept the invitation of Prof. Ernst-Joachim Mestmäcker, director of the Max Planck Institute for Foreign and International Private Law in Hamburg, to undertake the present study.

My focus is on INTELSAT as an international organization, not as one of many organizations on the international telecommunications scene. Furthermore, I have tried to strike a balance between describing the forest and the trees in that particular landscape. This has been more difficult than one might suppose. The reason primarily has to do with the juncture at which this study appears.

Early in 1987 I emerged from the geographic and, let us admit, the intellectual seclusion of Honolulu to spend a week in Washington, D.C. interviewing key individuals and gathering materials for this book. I found it difficult to induce the bright, able people with whom I spoke – whether in law firms, among INTELSAT's executive staff, in the U.S. government or elsewhere – to focus on much more than the very immediate challenges facing INTELSAT, above all, the prospect of competition with INTELSAT by private satellites and fiber-optic cables.

I agree that these are the major destabilizing forces with which INTELSAT must reckon in the near term, and I have accordingly given them the most detailed coverage in Chapter IV, which examines INTELSAT's present challenges. Today, however, there is a certain ennui with which one contemplates too single-mindedly an emphasis on these two issues, which since 1983 have been talked nearly to death (but by no means studied to death!) by INTELSAT, private firms, lobbyists, lawyers and INTELSAT's member-countries, in particular, the United States government. Indeed, as I have suggested in this volume, much of the frustration in examining these issues is related to the very high ratio of talk to study. As an economist, for example, I have been struck by how little hard economic analysis has taken place with respect to what is first and foremost an economic issue – whether separate international systems will inflict »significant economic harm« on INTELSAT, in the fuzzy phraseology of Article XIV(d) of INTELSAT's intergovernmental agreement. My approach has been to suggest some of the basic conceptual issues both for and against separate entry and to call the reader's attention to what limited theoretical and empirical research, including some of my own, has been conducted to address the matter of economic harm. This being done, one has a feeling that many of these issues are now moot and that political factors have taken the upper hand in determining INTELSAT's evolution from adamant opposition to resigned acquiescence. It bears noting, however, that virtually all of the people with whom I spoke on the subject regard fiber-optic cables as a greater long-term economic threat to INTELSAT than separate satellite systems.

As a glance at the table of contents will indicate, this study examines INTELSAT's past and future as well as its admittedly troubled present. If anything, this broader perspective is a more reassuring one. INTELSAT will continue to operate in any event, and if it capitalizes on its experience and comparative advantage, it can prevail and thrive well into the coming century. This best-case scenario, however, will occur only if certain short-term sacrifices are made. I have argued that the adoption of greater pricing flexibility and the acceptance of facilities competition from other international and regional systems, both private and public, are high on the list. The successful Article XIV(d) coordination of the PanAmSat system in April 1987 is an indication that INTELSAT is now examining and adopting some of these options realistically and that the rhetoric on both sides has mercifully subsided.

In the first chapter I have provided non-specialist readers with an extended tutorial on the technical and economic issues necessary to evaluate INTELSAT and the challenges it faces. Chapter II is a thorough review of

INTELSAT's interim and permanent agreements as they relate to these issues. Chapter III samples INTELSAT's present policies and practices in the areas of tariff policy, the peaceful use of outer space, and access to the orbit and spectrum. Chapter IV, the heart of the study, draws on the material in the preceding chapters to examine seven key issues facing INTELSAT today. Chapter V looks both to the past, in suggesting some frameworks within which INTELSAT's experience to date can be analyzed and evaluated, and to the future, in projecting a number of ways in which INTELSAT can maximize its chances for a vibrant existence well into the twenty-first century.

I should like to thank a number of people who have been generous with their time and experience relating to INTELSAT. All of them spoke with me candidly under the agreement that I would not cite their opinions directly. The views I have expressed in this study, as well as the methodological and historical interpretations I have offered are entirely my own. They do not reflect the official position of INTELSAT, the Max Planck Institute, the United States government or any other public or private entity. For their freely provided views and suggestions, I am most indebted to Robert Atkins, Lewis Cramer, Michael Cummins, Robert M. Frieden, Henry Goldberg, John D. Hampton, Ellen D. Hoff, Francis Latapie, Timothy J. Logue, Larry F. Martinez, Tedson J. Meyers, Dean Olmstead, Joseph N. Pelton, William Schultz, Ivan Shefrin, and Phillip L. Spector. Finally, I wish to thank Professor Mestmäcker and his institute for making this study possible.

Marcellus S. Snow
Honolulu, April 1987

Table of Contents

Chapter IV: Major Issues Affecting Relationships Between INTELSAT and Its Member-Countries 83

List of Acronyms and Abbreviations

AID	Agency for International Development (U.S.)
AT&T	American Telephone & Telegraph Corporation (U.S.)
BT	British Telecommunications (U.K.)
CCIR	International Radio Consultative Committee (ITU)
CCITT	International Telephone & Telegraph Consultative Committee (ITU)
Comsat	Communications Satellite Corporation (U.S.)
DAMA	Demand assigned multiple access
ESA	European Space Agency
ESOC	Earth Station Owners Consortium (U.S.)
EURATOM	European Atomic Energy Commission
FCC	Federal Communications Commission (U.S.)
FDM	Frequency division multiplex
FM	Frequency modulation
G/T	Gain-to-noise temperature ratio
IADP	INTELSAT Assistance and Development Program
IBI	International Bureau for Informatics
IBS	INTELSAT Business Service
ICAO	International Civil Aviation Organization
ICSC	Interim Communications Satellite Committee (INTELSAT)
IDF	INTELSAT Development Fund
IFRB	International Frequency Registration Board (ITU)
IIC	International Institute of Communications
IMCO	International Maritime Consultative Organization
INMARSAT	International Maritime Satellite Organization
INTELSAT	International Telecommunications Satellite Organization
ISDN	Integrated switched digital network
ITU	International Telecommunication Union
kbps	Kilobits per second
KDD	Kokusai Denshin Denwa Co., Ltd. (Japan)
LATA	Local access and transport area (U.S.)
LDC	Less-developed country
MITI	Ministry of International Trade and Industry (Japan)
MOU	Memorandum of understanding
MPT	Ministry of Posts and Telecommunications (Japan)
NASA	National Aeronautics and Space Administration (U.S.)
NTIA	National Telecommunications and Information Administration (U.S.)
NTT	Nippon Telegraph and Telephone Corporation (Japan)
OECD	Organization for Economic Cooperation and Development
OFTEL	Office of Telecommunications (U.K.)
OPEC	Organization of Petroleum Exporting Countries
PCM	Pulse-code modulation
PSK	Phase shiff keying
PTC	Pacific Telecommunications Council
RITE	Research Institute for Telecommunications and Economics (Japan)
SHARE	Satellites for Health and Rural Education (INTELSAT)
WARC	World Administrative Radio Conference (ITU)

Chapter I: Historical, Technological and Economic Background

A. *INTELSAT's Origin and Development*

As an international organization the International Telecommunications Satellite Organization – INTELSAT – is unique in combining the diplomatic machinery of an intergovernmental body with a commercial, operational mission: that of providing the capacity required for more than half of all international telecommunications links. Initially shielded from severe internal political tensions by its concentration on a technological and operational mandate, INTELSAT has found, as it enters its third decade, that the very dynamism of the technology it provides has introduced severe political and economic challenges among its 113 members. This study is an account of how and why those challenges came about and what they portend for the future of INTELSAT, international telecommunications and the users and providers of satellite communications.

INTELSAT was born in a force field of technology and international politics reflecting conditions in the late 1950s and early 1960s.[1] As early as 1945, Arthur Clarke (1945) had recognized that orbiting artificial satellites could act as »radio relays« in receiving, amplifying and retransmitting electromagnetic signals between points on earth. In addition, using physical principles familiar for centuries, he calculated the altitude above the earth's equator – approximately 36,000 km – at which gravitational forces are such that an orbiting object rotates with the same angular velocity as the earth and therefore appears to hover in place above a single point.

The first Soviet and U.S. satellites, launched in 1957 and 1958, respectively, were small, crude repeater devices emplaced in subsynchronous orbit. By the early 1960s, experiments with the U.S. Echo and Telstar satellites, as well as initial success in reaching the geostationary orbit, convinced the major Western countries of the commercial promise of satellites for international public telecommunications. The institutional arrangements arrived at testify to the geopolitical realities of the Cold War and to the uneven distribution of satellite and launcher technology that then prevailed.

1 Early treatments of INTELSAT and its origins include those of Galloway (1972), Kildow (1973), Pelton (1974) and Snow (1976).

The initial focus was naturally on the industrialized countries on either side of the Atlantic wishing to use satellites to supplement their undersea cables and high-frequency radio links already in operation. The Soviet Union declined to join the nascent organization and instead created another international body, INTERSPUTNIK.[2] The United States, Canada, most of Western Europe, Japan, Australia and several other countries then proceeded to negotiate international institutional arrangements enabling them to introduce commercial public satellite communications across the Atlantic and, eventually, on a global basis as well.

In the early 1960s, however, the United States held the upper hand in technological prowess and political influence. Thus, the INTELSAT organization originally created by treaty in 1964 reflected that reality (see details in Chapter II, Part A). In particular, initial INTELSAT operation was carried out on the Early Bird Satellite, a U.S. spacecraft rechristened as the INTELSAT I. The second generation INTELSAT II satellites depended to a considerable extent on traffic from the National Aeronautics and Space Administration (NASA), the U.S. space agency, for their financial viability. Finally, INTELSAT was completely dependent on NASA for synchronous launch services throughout the 1960s and 1970s.

The major challenges facing INTELSAT in the late 1960s – by which time it was operating third-generation satellites in the Atlantic, Pacific and Indian Ocean regions – were (1) a transition to a permanent regime embodied in renegotiated, definitive treaties; (2) increased opportunities for Western European aerospace industries to provide subcontracting for INTELSAT satellites and services; (3) a decrease generally in the influence of the United States government in INTELSAT; and (4) a curtailment of the role and activities of the U.S. signatory to INTELSAT, the Communications Satellite Corporation (Comsat). These goals were by and large achieved by the signing of the renegotiated INTELSAT agreements, which entered into force in February 1973, and by substantial gains in non-U.S. subcontracting for the INTELSAT IV and subsequent satellite series.

A number of longer-term concerns, many of which persist to the present day, began to manifest themselves in the early 1970s. In essence, patterns of technology, economics and user demand challenged INTELSAT's hegemony as the predominant supplier of public international satellite services. First, the

2 The origin of INTERSPUTNIK is discussed in Galloway (1972, Chapter 7) and Shillinglaw (1970). Soviet telecommunications productivity is estimated by Lewis (1975). Today INTERSPUTNIK is essentially an extension of the Soviet domestic satellite system to that country's political allies in Eastern Europe, the Caribbean, Africa and elsewhere.

International Maritime Satellite Organization (INMARSAT) was established in the mid-1970s – and indeed was consciously modeled after INTELSAT – to provide communications and navigation by satellite for ships at sea. Second, various domestic and regional satellite systems began to emerge. Canada and Indonesia made massive and early commitments to national systems, while the United States, after years of deliberation, adopted in 1970 an »open skies« policy[3] to allow the market-driven evolution of multiple and competing domestic systems. Regional systems, particularly in Europe and among the Arab countries, were in place by the late 1970s. INTELSAT itself, seeking to maintain its global role and to forestall the proliferation of such separate systems, began in 1973 to permit some of its long-standing excess capacity to be used at concessionary rates by its members on an interruptible basis for domestic applications.
The 1980s brought two additional challenges to INTELSAT. First, its less-developed members, which had through the 1960s and 1970s seen their interests as having been served by the United States and by a major U.S. influence in INTELSAT, began to exert pressure for policies more attuned to the needs of the developing world. These included in particular small, less-expensive earth stations and more flexible and concessionary tariff policies. INTELSAT replied with a variety of measures (see Section G of Chapter IV). Second, INTELSAT saw its preeminence in international communications threatened by the advent of private satellite systems and by the technological breakthrough of newly developed fiber-optic cables. These challenges in turn introduced centrifugal political and economic tensions into an international organization that had for two decades enjoyed a relatively apolitical existence as it evolved toward its present mature configuration. The dynamics of these forces are played out in particular in the changing pattern of interaction between INTELSAT and its member-countries, which, as a major theme of this study, is surveyed in detail in Chapter IV.

B. *Technological Background*

1. *The geostationary orbit*

As Arthur Clarke reminded the world in 1945, the laws of physics provide an orbit 36,000 km above the equator in which objects appear from earth to

3 See Nelson (1977) for an account of the origin of U.S. domestic satellite policy.

hover in equipoise. This greatly decreases the cost of the earth segment of a synchronous satellite system, since ground stations do not need the expensive equipment required to track subsynchronous satellites across the sky several times a day. Thus, the geostationary orbit – often called the Clarke orbit – is a precious natural resource akin in many ways to the more familiar radiofrequency spectrum. Its value derives not only from its limited physical extent but also from the great expense involved in reaching it by appropriate rocketry. There are, to be sure, drawbacks of the geostationary orbit. First, it is too high to be useful for certain non-communications satellite applications like remote sensing. Such activities must be carried out by subsynchronous spacecraft orbiting only a few hundred kilometers above the earth. Second, the remoteness of the geostationary orbit causes a delay of approximately one-quarter of a second in transmitting a message there and back. This is perceived as objectionable by some telephone users and presents difficulties for various data transmission applications as well.

2. *Launchers*

The earliest satellites, despite their limited size and mass, could be launched only into subsynchronous orbit. Indeed, the initial capitalization of the Communications Satellite Corporation assumed subsynchronous technology and was several million dollars too high as a result. By 1965, however, when INTELSAT began operation, NASA was able to place the organization's first-generation spacecraft into synchronous orbit with a Thor Delta launcher, which also provided launch services for the INTELSAT II and III series. The larger INTELSAT IV and IV-A satellites were launched aboard NASA's Atlas Centaur rockets, while the current INTELSAT V and V-A series have been placed into orbit aboard both Atlas Centaur and Ariane 1 and 2 rockets, the latter furnished by the Arianespace enterprise of the European Space Agency (ESA) at a launch site in French Guiana. Both the reusable NASA space shuttle and the Ariane 4 expendable rocket are envisaged for the INTELSAT VI series to begin operation by 1989, but launch accidents suffered by both organizations in early 1986 have postponed the future availability of these or similar vehicles. Launch availability, as a result, has surprisingly become one of the major imponderables facing all satellite systems during the next five years or so.

Recent launches by INTELSAT and other satellite operators have been insured by large consortia of insurance firms. Some operators have been reimbursed upon failure of their spacecraft to reach and operate in geostationary orbit; others are engaged in litigation to collect compensation from insurers under conditions in which the extent of launch success is contested. The terms and prices of launch insurance have varied quite unpredictably, and there is little evidence that the insurance industry yet fully understands the nature and risks of the associated technology or the appropriate principles by which to price launch insurance policies.

Table 1 – INTELSAT Financial Data, 1964-85
(All Figures in Millions of U.S. Dollars)

Year	Depreciation	Operating Expenses	Total Expenses	Total Revenues	Capital Expenditures	Compensation for Use of Capital
1964	0	3.9	4.0	0	3.8	0
1965	3.0	4.9	8.7	2.3	10.8	0
1966	7.6	4.0	12.5	3.3	23.5	0
1967	19.2	5.2	24.4	16.7	41.3	-7.7
1968	13.6	7.6	21.2	28.9	30.1	7.7
1969	23.5	12.0	35.4	44.7	60.2	9.2
1970	18.3	13.3	31.6	75.4	87.3	43.7
1971	27.9	14.7	42.6	78.3	68.5	35.6
1972	38.9	15.7	54.6	89.6	42.7	35.0
1973	43.8	17.3	61.1	99.3	65.9	38.8
1974	39.2	18.3	57.5	101.4	72.5	44.0
1975	42.2	22.8	65.0	117.7	85.8	52.8
1976	55.1	27.0	82.2	143.4	71.8	61.2
1977	57.8	30.4	88.2	157.3	105.5	69.1
1978	70.0	34.6	104.6	185.9	107.6	81.3
1979	63.5	35.7	99.2	194.9	144.1	95.7
1980	55.7	46.6	102.3	216.3	201.5	114.0
1981	75.1	56.4	131.5	249.1	254.4	117.6
1982	89.3	66.5	155.8	315.4	331.8	159.6
1983	100.7	76.7	177.4	366.2	397.1	188.8
1984	115.1	87.6	202.8	411.3	258.9	208.4
1985	129.2	103.3	232.5	458.3	250.9	224.1

Sources: INTELSAT Annual Reports and Annual Financial Reports, 1964-85

3. *Satellites*[4]

Tables 1 through 3 indicate that INTELSAT, during more than twenty years of operation, has employed satellites with exponentially increasing levels of capacity, physical size and technological sophistication. The initial generations

4 Podraczky and Pelton (1984) give an overview of INTELSAT satellite technology. A more general summary as of the mid-1970s is that of Bargellini (1974).

Table 2 – INTELSAT Financial and Operating Data

Year	Return on Use of Capital (%) Year*	Cumulative*	Depreciated Value of Capital Stock at Year's End (US $000,000)	Full-time Circuits at Year's End	Annual Charge for Half-Circuit (US $)
1964	--	--	7.8	--	--
1965	--	--	21.9	75	32,000
1966	--	--	47.1	86	20,000
1967	-11.8	-11.8	69.3	344	20,000
1968	9.6	0.0	85.8	571	20,000
1969	9.0	3.7	122.5	1,417	20,000
1970	27.5	13.0	191.5	2,129	20,000
1971	16.5	14.2	232.1	2,911	15,000
1972	14.5	14.3	235.9	3,748	12,960
1973	15.5	16.9	258.0	4,907	11,160
1974	14.7	15.6	291.4	5,753	9,000
1975	14.8	15.3	335.0	6,684	8,460
1976	15.7	15.4	351.6	8,260	8,280
1977	16.0	15.6	399.3	10,100	7,380
1978	16.8	15.9	436.9	12,647	6,840
1979	17.8	16.2	523.5	16,209	5,760
1980	17.7	16.5	672.5	20,308	5,040
1981	14.5	16.1	852.0	25,133	4,680
1982	15.9	16.1	1,094.5	29,737	4,680
1983	14.3	15.7	1,390.8	32,731	4,680
1984	13.5	15.3	1,534.5	36,278	4,680
1985	13.6	15.5	1,640.4	40,335	4,680

* The 1967-72 figures are cumulative from 1 January 1967. The 1973-85 figures are cumulative from 12 February 1973. The $19.3 million revenue shortfall for 1964-66 was capitalized over the 10-year period 1967-76 as a system development cost. 12 February 1973 was the date of entry into force of the permanent INTELSAT agreements.

Sources: INTELSAT Annual Reports and Annual Financial Reports, 1964-85

partook to some extent of earlier design work on U.S. military satellites, particularly by Hughes Aircraft, and thereby can be considered as having embodied a certain subsidy by U.S. taxpayers. Later INTELSAT series incorporated various technological innovations, such as spin stabilization and spot beams.

The mass and communications capacity of a satellite figure in at least two fundamental tradeoffs involving system-wide economic and engineering optimization. The first concerns the relationship between satellite mass and

Table 3 – Specifications of INTELSAT Satellites, 1965-87

Series	First Launch	Prime Contractor	Width (m)	Height (m)	Launch Vehicle	Design Lifetime (Yrs.)	Bandwidth (MHz)	Capacity In Voice Circuits
I	1965	Hughes	0.7	0.6	Thor Delta	1.5	50	240
II	1967	Hughes	1.4	0.7	Thor Delta	3.0	130	240
III	1968	TRW	1.4	1.0	Thor Delta	5.0	300	1,500
IV	1971	Hughes	2.4	5.3	Atlas Centaur	7.0	500	4,000
IV-A	1975	Hughes	2.4	6.8	Atlas Centaur	7.0	800	6,000
V	1980	Ford Aerospace	2.0	6.4	Atlas Centaur Ariane 1,2	7.0	2,144	12,000
V-A	1985	Ford Aerospace	2.0	6.4	Atlas Centaur Ariane 2	7.0	2,250	15,000
VI	1987	Hughes	3.6	6.4	NASA STS Ariane 4	10.0	3,300	30,000

Source: INTELSAT Annual Report, 1985-86

launcher power. Is it more economical to orbit a larger number of lighter satellites using smaller rockets or a smaller number of heavier spacecraft requiring more powerful launchers? The second tradeoff is between earth stations and satellites. Basically, the communications signal can be amplified both at the transmitting and receiving earth station and by the satellite itself. Is it advisable to launch a more powerful (and therefore heavier and more costly) satellite, requiring a larger launch vehicle, that can operate with smaller earth stations required to perform less signal amplification? If the converse is true, then the use of a small, low-powered satellite that must work with larger, more expensive earth stations capable of greater signal amplification is indicated. There has been a gradual evolution from the former to the latter regime during INTELSAT's history, one that has taken greater account of the needs of developing countries and has not been without political implications among the organization's member-countries.

4. *Earth stations*[5]

The major problem in effecting a system-wide engineering and economic optimization of INTELSAT's space segment with its associated earth stations is that the latter are not part of INTELSAT but are instead owned and operated either by the individual signatories to INTELSAT's agreements or by third parties. Legal, institutional and political considerations militated against incorporating the earth segment into INTELSAT's ownership and administrative structure. Thus, complete system optimization between the single satellite operator and some 100 administrations accessing the system by means of their own terminals is effectively impossible.
The technology of the early 1960s prescribed small satellites, the only kind light enough to be launched into geostationary orbit with the rockets then available. This in turn dictated large earth terminals capable of greatly amplifying the weak signal received and transmitted by the satellites. The initial standard earth station prescribed by INTELSAT had parabolic dishes with a diameter of some thirty meters that cost several million dollars each. The size and expense of the early satellites made them accessible only to the industrialized members of INTELSAT and to the handful of larger developing nations able to spare the foreign exchange necessary to purchase them. In addition, the use of one or a small number of large stations, as opposed to many smaller ones, placed a greater interconnection burden on the existing terrestrial infrastructure of the user countries. As noted, more powerful satellites, as well as pressure on INTELSAT by its developing country members, are in the process of bringing about a slow evolution toward smaller, cheaper earth stations better suited to the needs of most developing countries.

5. *Tracking, telemetry and command*

INTELSAT's assets and operational mission also comprise equipment for tracking, telemetry and command of in-orbit satellites. These facilities are located in INTELSAT's main control room at its Washington, D.C. headquarters as well as at various special earth stations around the world that contract with INTELSAT to carry out these functions. Station keeping of orbiting satellites – ensuring that they remain at given longitudinal locations above the equator – is a costly and demanding task. In addition, satellites are

5 See Nosaka (1984) for a survey of earth station technology.

occasionally drifted between locations over a given ocean region, or even from one region to another, as they are replaced by newer spacecraft or as other operational requirements dictate. These drifting maneuvers, performed by firing jets that use on-board fuel, are a constraint on a satellite's useful lifetime, since weight requirements permit only a limited amount of fuel to be carried into orbit.

Since earth-station costs have risen moderately while the cost of space-segment capacity has fallen exponentially, this has meant that the cost share of an end-to-end satellite connection actually attributable to the space segment has steadily decreased. It is presently in the neighborhood of ten percent and still falling. It is this evolution that has been primarily responsible for the increased attention to earth-station costs and configurations by INTELSAT users, especially those in the Third World.

C. *Economic Issues*[6]

1. *Cost concepts and structure*

Costs and technology are captured by the cost function, which is a mathematical relationship specifying the cost of output as a function of output levels and input prices. Two important properties of cost functions bear mention here, one of which is static (abstracting from the passage of time) and one of which is dynamic (taking the passage of time explicitly into account). First, a given cost function may exhibit economies of large-scale operation in the sense that the average cost of higher levels of output is less than it is for lower output levels. This static property is usually called economies of scale. By contrast, the parameters and perhaps the mathematical form of the cost function itself may change over time, making the average cost of *any* output level fall. This is the dynamic phenomenon of technological change. If economies of scale are more important than technological change in satellite operation, then research should be directed toward building higher-capacity spacecraft within the existing technology. Such economies, to be sure, are generally exhausted beyond certain output levels. If technological change is the greater source of cost savings, however, research should focus on innovation and adaptation of the technology rather than on output scale alone.

6 For a more complete overview of satellite economics, see Snow (1987a). A broad introduction to telecommunications economics is that of Wenders (1987).

There are a number of distinguishing features in the cost structure of communications satellite systems. The most important is the insensitivity of satellite costs to the terrestrial distance of transmission, at least within the coverage zone of a single satellite. For a global beam from synchronous orbit, this zone encompasses approximately one-third of the earth's surface, excluding the polar regions. The distance insensitivity of satellites costs – given the cost sensitivity to distance of earth-bound transmission modes – implies that the economic advantage of satellites over terrestrial modes of transmission will be greatest when long distances, mountainous terrain, severe climate and sparsely populated regions separate the sender from the receiver.

In addition, competing modes, such as cables, require costly hierarchical switching to achieve interconnectivity. The costs of switching components have risen considerably in recent years due to the extensive research and software that were required with the advent of digital switching and other innovations.

Very little effort has been devoted to estimating cost relationships for INTELSAT. Initially, Snow (1976, pp. 19–21) used data from the first decade of INTELSAT operation to estimate a cost function relating a single output variable – capacity in equivalent voice-grade circuits – to total cost in dollars. In doing so, he found strong evidence of static scale economies.

Two consulting studies have been commissioned in more recent years by interested parties in the debate concerning private satellite systems in competition with INTELSAT. Walter Hinchman Associates (1984), consulting for INTELSAT, concluded that the organization's cost structure was such that it would be harmed by competition, while Dale L. Hatfield Associates (1983), consulting for a prospective entrant, reached the opposite conclusion. Both of these studies, however, used an accounting rather than an econometric approach to the allocation of common costs, implying a certain degree of arbitrariness.

Snow (1987b), using econometric techniques and eighteen years of financial and operating data, has recently estimated a multioutput cost function for INTELSAT by specifying a transcendental logarithmic (translog) cost function. He addressed the question of whether INTELSAT exhibits the property of natural monopoly, a subject that will be dealt with in Section 6(a) below. His results support the hypothesis of natural monopoly in INTELSAT. It is of interest that Evans and Heckman (1983a, 1983b, 1984), using the same methodology with data on the pre-divestiture American Telephone and Telegraph Company (AT&T), arrived at precisely the opposite conclusion, namely, that the Bell System did not constitute a natural monopoly. One policy import of these findings might be that the agreement allowing the divestiture of AT&T was in the public interest but separate systems competing with INTELSAT would not be. Snow's study, nevertheless, should be seen as a tentative effort subject to numerous qualifications, including lack of primary data. Curiously, however, it is the only extensive econometric cost study of INTELSAT to date, despite the ostensibly economic character of the debate surrounding the entry of separate systems into competition with INTELSAT. This issue will be taken up in detail in Section A of Chapter IV.

2. *Demand concepts and patterns*[7]

Since they relate to a sector of the economy that has traditionally been furnished in most countries on a monopoly basis by government agencies or by regulated private firms, telecommunications systems and planning have most often reflected the values and priorities of what might be called the »supply side.« The recent advent of widespread liberalization in the telecommunications sectors of industrialized countries, however, has accorded more influence to telecommunications users – the »demand side.« These transformations have included the privatization of domestic and international telecommunications providers and the introduction of competition with the dominant telecommunications carrier, usually a former monopoly organization under government ownership, control and operation. Until now, user pressure has come primarily from large multinational corporations depending on international telecommunications networks, such as those providing worldwide airline reservations and other financial transactions. More recently, smaller users, including residential consumers, have become influential, particularly in the political arena.

Demand for telecommunications and other services is affected by a number of factors. Economically, the most important such determinants are the price of the services and the income level of the user. Econometric equations estimated to determine the strength, direction and functional form of these relationships are called demand functions. The effect of price on determining the quantity demanded is measured by the price elasticity of demand, which is the absolute value of the percentage change in demand brought about by a one-percent change in price. Likewise, income elasticity of demand is the percentage change in demand caused by a one-percent change in income.

7 See Taylor (1980, 1983) for a general discussion of demand for telecommunications. Rea and Lage (1978) and Corbo and Munasinghe (1978) give empirical estimates of demand elasticities for telecommunications.

Numerous studies have confirmed that the price of telecommunications services, and the income of telecommunications users, significantly affect the demand for services. Nevertheless, the projections of telecommunications demand used by planners to determine requirements for new facilities often neglect demand relationships, or assume them to be insignificant. INTELSAT's chronic problem of excess capacity (see Johnson 1986) – a difficulty familiar in many other telecommunications settings as well – can be traced in part to the neglect of economic factors, such as price, income, complementarity and substitution, in forecasting demand and the facilities required to satisfy that demand.

Demand elasticities also play an important role in determining pricing strategies and overall tariff structures, a topic to which we now turn.

3. *Tariff concepts and issues*

In competitive sectors of the economy in which neither buyers nor sellers possess market power – that is, the ability to affect price – the market autonomously creates incentives that drive prices toward costs. In short-term equilibrium, price will equal marginal cost, and in the long run, it will also equal average cost, which reaches a minimum when it intersects with marginal cost.

Under the monopoly provision of services that has historically obtained in the telecommunications sector, conditions have exerted less pressure to equate costs to prices. Private monopoly providers, such as licensed common carriers in the United States and Canada, have been constrained by regulatory authorities to charge prices that yield a profit on a »rate base« of investment that is close to the return available in closely allied competitive industries. Such rate-base regulation[8] ultimately assures an overall correspondence of prices with the underlying cost structure, although profits from remunerative monopoly traffic are sometimes used to subsidize below-cost tariffs on routes subject to competitive entry. In addition, public monopolies are often under severe political pressure to allow prices to deviate from costs, generally in order to provide internal or external subsidies to politically powerful and favored groups. As we shall see in Section A of Chapter III, INTELSAT is constrained by its agreements to an overall regime of average-cost pricing, although this as a practical matter has not prevented it from engaging in more

8 Snow and Jussawalla (1986, Chapters 2 and 3) survey and annotate the economics literature relating to telecommunications regulation.

innovative forms of tariff policy, including demand-based pricing, for particular services.

Indeed, demand-based pricing, which is in principle not possible under competitive market regimes, has certain properties of economic welfare optimality when monopoly conditions prevail. For example, Neumann, Schweizer and von Weizsäcker (1983) have shown that a welfare gain on the order of DM 1 billion would have been realized if the Deutsche Bundespost, West Germany's telecommunications authority, had engaged in demand-based pricing rather than in politically determined tariffing during the single year 1978. Nevertheless, as one moves from a static to a dynamic framework, and from monopoly to competition in the provision of services, the optimal tariff regime is based on cost rather than on demand (see von Weizsäcker 1986).

4. *Capacity planning and utilization*[9]

Satellite capacity, like other modes of telecommunications, must be provided in large, discrete increments rather than continuously. Since demand tends to grow in small amounts, however, a planning choice must be made between building capacity ahead of demand and allowing capacity to fall behind demand and then catching up.

Such a problem can be conceptualized in the following framework. Suppose that demand for satellite capacity grows exponentially at a known rate and that there are economies of scale in providing such capacity. Imagine that one wishes to satisfy demand for capacity indefinitely into the future and that the investment costs of doing so are discounted to the present at a constant annual rate and are summed over an infinite time horizon. There must then be periodic »regeneration points« at which growing demand catches up with supply and a new satellite must be launched. The investment-cost minimization problem can now be stated as follows: how large a satellite should be launched? Stated differently, for how many years into the future should demand be anticipated? The tradeoff is apparent. Launching too small a satellite – selecting too brief a period between regeneration points – would reduce the effects of economies of scale. On the other hand, orbiting too large a spacecraft – satisfying demand too far into the future – would lead to excess capacity and shift too much of the cost into today's budget instead of moving it (appropriately discounted) into later periods. Intuition suggests that there

9 Schnicke (1984 A&P) summarizes INTELSAT's system planning practices and philosophy.

should be some optimal spacing of satellite launches over time (regeneration points) such that discounted investment costs are minimized over an indefinite time horizon. Srinivasan (1967) greatly simplified this problem by showing that under quite general conditions, cost minimization implies that regeneration points are equally spaced.

Snow (1975) augmented this model in two ways and then applied it to INTELSAT operating data. First, he incorporated the assumption of exponentially depreciating in-orbit capacity into the model to replace the unrealistic implication that capacity once in orbit would never become inoperative. Second, he showed how the effect of price on demand could be accommodated in a mathematically tractable fashion. Within the context of the extended model, it was found that the theoretically indicated optimal capacities of INTELSAT spacecraft, as well as the optimal regeneration interval between launches, were not too different from the capacities, design lifetimes and launch patterns actually used by INTELSAT during its first decade of operation. This implied, in turn, that INTELSAT had pursued cost minimization as a conscious goal, a policy that benefited both INTELSAT owners and users, given the cooperative structure of the organization.

As a practical matter, plans and projections for INTELSAT utilization are submitted by INTELSAT's users to periodic meetings of operational representatives. These data, in turn, are the foundation on which INTELSAT bases its planning for future capacity and launches. Snow (1976, pp. 133–34) and Johnson (1986) have suggested reasons for overoptimistic usage projections that have been responsible in large part for the excess capacity that has historically been associated with INTELSAT.

5. *Cost, prices, demand and capacity: a unifying framework*

Subsections 1 through 4 immediately above each outline an important aspect of economic reality for INTELSAT or, indeed, for any telecommunications system. Figure 1 represents the most obvious links among these four variables; these relationships are circular, in the sense that each component is both the result and the cause of another.

Let us begin with capacity, which is related to cost by means of the cost function. This is the connection that encompasses the state of the technology being used, as well as market conditions bearing on input prices. Cost, in turn, is related to prices through a tariff policy. As we have seen, prices can bear a rather arbitrary relationship to cost if they are determined politically within a monopolistic supply framework. In a competitive setting, market pressures

Figure 1

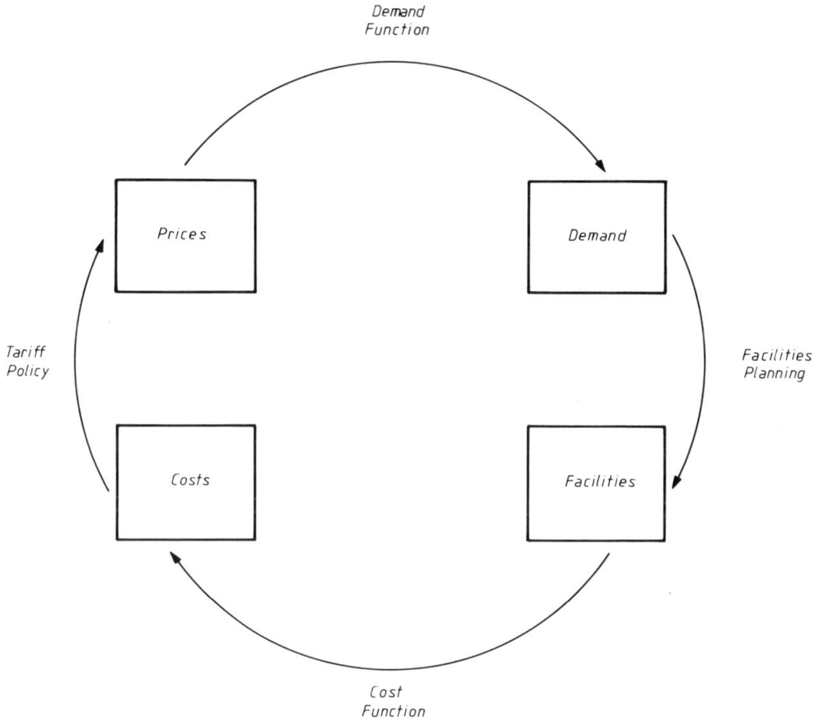

automatically force prices toward costs. Prices, in turn, are the main determinants of demand, although exogenous causes such as income can also affect demand. It is the demand function that relates prices and other variables to the level of demand. This autonomous behavioral relationship is the one most often ignored by telecommunications planners, who have habitually used exponential growth projections for demand independently of any assumptions about price or its effect on demand. Finally, demand is related to capacity by the system planning process. This is the relationship over which the greatest influence can be wielded by planners and in which the greatest arbitrary element inheres. Once capacity is determined, however – perhaps by a cost-minimization model of the kind set forth in Subsection 4 above – the circle of causation is closed, and capacity once again gives rise to costs through what is possibly a new cost function in a planning period.

Figure 1 is merely schematic and ignores many other relationships. Costs are also affected by the prices of inputs, for example, which are exogenous to the model. In addition, outside political and social factors may affect the tariff policy agreed upon. Despite its oversimplification and abstractness, Figure 1 is nonetheless well beyond the degree of analytical sophistication presently employed in most telecommunications systems. Such systems generally approach these four relationships in a disjointed fashion, using different departments and personnel. Scientists and engineers typically handle the cost function; economists, accountants and lawyers are in charge of tariff policy; marketing experts examine the demand function; and demand is accommodated to capacity by system planners and operational representatives. Problems such as excess capacity and internal subsidization of services, for example, are evidence that these relationships are studied in virtual isolation. A challenge to INTELSAT and other telecommunications providers is to integrate these financial, technological and behavioral relationships with an eye to greater system efficiency as competition from other systems increases.

6. *Recent industrial organization theory relevant to space telecommunications*

Our survey of economic concepts and issues underlying satellite communications concludes with a brief tutorial setting forth the growing literature on natural monopoly, sustainability and contestability of market structure[10] in the economic subdiscipline of industrial organization. These concepts have been extremely useful in arriving at novel and insightful conceptualizations of satellite communications and other aspects of the telecommunications sector; indeed, their origin derives in large part from the pressing need to develop new theoretical perspectives capable of analyzing the technological, regulatory

10 The most rigorous and encyclopedic reference to these related topics is that of Baumol, Panzar and Willig (1982), which is reviewed by Brock (1983) and Spence (1983). Bailey and Friedlaender (1982) furnish a useful and extended bibliographic essay on the same subject matter. Sharkey's (1982) treatment of natural monopoly and sustainability and Baumol's (1982) address on contestability theory to the American Economic Association are perhaps the most straightforward and accessible introductions to this literature. Shepherd's (1983, 1984) pieces are indicative of an opposing point of view, while Coursey, Isaac and Smith (1984) present empirical evidence confirming the behavioral predictions of the contestable-market hypothesis. Baumol, Panzar and Willig (1986) present new results on contestability and discuss in particular the empirically realistic case of small but positive sunk costs. Bailey and Baumol (1984) consider contestability in the context of U.S. deregulatory trends. Snow and Jussawalla (1986, Chapter 6) survey and annotate the literature on natural monopoly, sustainability and contestability.

and economic challenges of the entire information sector of today's global economy. Each of the headings below describes one important aspect of this new complex of theory and gives examples of how it might be or already has been applied to conceptual, methodological and policy problems affecting INTELSAT.

a. *Natural monopoly*

It is useful to distinguish the de jure concept of legal or administrative monopoly from the de facto notion of natural monopoly. The latter property is purely economic and technological; mathematically, it can be expressed and estimated as a property of the cost function.

Conventional microeconomic theory posits that all competitors in a given market produce a single identical commodity. Under this assumption, the existence of economies of scale (declining average costs) at least to the level of market output is sufficient to ensure natural monopoly and to deter competitive entry. By producing at a higher level of output, the incumbent is able to realize lower costs and thereby to offer lower prices than can any competitor or group of competitors producing at lower levels and thereby incurring higher costs. Indeed, the ability of a natural monopolist to repel profitable entry by exploiting economies of scale gave rise to the extensive theory of regulation familiar in the economic literature today. Thus, economies of scale and natural monopoly are essentially identical properties in a single-output framework.

Beginning in the mid-1970s, researchers expanded their framework of inquiry to include multiproduct outputs. As it turned out, this quantitative augmentation of the production function provided novel and often counterintuitive results. The most important of these is that a producer possessing natural monopoly is no longer necessarily protected against successful entry if two or more outputs by each producer are allowed for.

Whether or not a producer possesses a natural monopoly is an important empirical question with obvious ramifications for policy makers and regulatory authorities. While, as we shall see, the presence of natural monopoly does not always justify regulatory barriers to entry or outright prohibition of entry, the absence of natural monopoly is certainly a strong argument against such policies.

The property of natural monopoly can be expressed mathematically by means of relationships involving cost functions. The technology available to a group of firms in an industry (we assume that each firm has access to and uses the same technology) can be summarized by a cost function C( ), which indicates the cost C of producing at a certain level of output. If a single output is produced at level x, then the cost of doing so is C(x). For a richer analysis, we assume that there are n separate outputs, produced at levels $x_1, x_2,\ldots,x_n$, respectively. Then the cost of such a bundle of output can be expressed as $C(x_1,x_2,\ldots,x_n)$.

Suppose now that society can choose between allowing the output vector $(x_1,x_2,\ldots,x_n)$ to be produced either by a single firm or by two or more firms jointly. Let there be m such firms, where $m \geqslant 2$. The production cost for the single firm will be $C(x_1,x_2,\ldots,x_n)$. Let the output bundle of the ith firm be $(x_{li},x_{2i},\ldots,x_{ni})$, where $i = i,2,\ldots,m$ and $\sum_i x_{ji} = x_j$ for $j = 1,2,\ldots,n$. Some of the x_{ji} elements will be equal to zero in cases of specialized production. Since each firm has the same technology and cost function, firm i's output cost will be given by $C(x_{li},x_{2i},\ldots,x_{ni})$. For the aggregate output bundle in question, the single firm is said to exhibit natural monopoly if

$$C(x_1,x_2,\ldots,x_n) < C(x_{11},x_{21},\ldots,x_{n1}) + C(x_{12},x_{22},\ldots,x_{n2}) + \ldots + C(x_{1m},x_{2m},\ldots,x_{nm}).$$

In words, natural monopoly is present for a given output bundle if a single firm is able to produce that bundle at a lower cost than can two or more firms in concert. Global natural monopoly obtains if natural-monopoly holds for all possible output bundles. Natural monopoly as expressed above corresponds to what is called subadditivity of the cost function.

Since INTELSAT has been threatened with competitive entry on its Atlantic routes for several years, the question of whether is possesses natural-monopoly properties is an important one. Indeed, as we shall elaborate in Section A of Chapter IV, Article XIV(d) of the INTELSAT intergovernmental agreement prescribes that competing public international systems not inflict »significant economic harm« on INTELSAT.

As noted earlier in this chapter, Snow (1987b), using econometric estimation of a translog cost function, determined that INTELSAT possessed natural monopoly for each of sixty-five artifically constructed output disaggregations from 1976 through 1984. His cost function had as arguments two output measures, Atlantic and non-Atlantic circuits; two input price measures, for capital and labor; and an index of technological change. By way of illustration, one such disaggregation is the following. In 1984 INTELSAT alone provided 22,935 Atlantic circuits and 13,296 circuits elsewhere, at an econometrically estimated cost of $427.1 million. Suppose that two firms (one of which would presumably be INTELSAT) were to share this market in the following way. Firm A provides 5,838.25 circuits to the Atlantic and 3,391.50 circuits elsewhere. Firm B provides 17,096.75 circuits to the Atlantic and 9,904.50 circuits elsewhere. Extrapolating from past years in which INTELSAT's output was similar to these smaller amounts, and allowing for technological change in the interim, the costs of hypothetical Firm A are estimated at $391.7 million, and those of Firm B, at $415.1 million. Producing together, therefore, Firms A and B incur total costs of $806.8 million, which is 91.3 percent more than the $421.7 million it historically cost INTELSAT to produce the entire output by itself.

There are, to be sure, countervailing considerations. Natural-monopoly theory is set in a static mold and does not take into account the ability of competition to bring about technological change or new goods and services. Further, as noted, it assumes that the same technology is shared by all firms and that the incumbent acts passively in the face of competitive threats. Nonetheless, the presence of natural monopoly is certainly a necessary if not a sufficient condition to justify prohibition of entry.

b. *Sustainability of natural monopoly*

If there are two or more outputs involved, the property of natural monopoly no longer necessarily protects the incumbent against the profitable entry of rivals. If no such protection is afforded, the monopoly is said to be unsustainable. Otherwise it is called sustainable.

While natural monopoly is a property of the cost function alone – which in turn relates to input prices and the underlying technology – sustainability of natural monopoly depends in large part on the pricing strategy used by the incumbent. Unfortunately, little is known about the conditions under which a natural monopoly can be rendered sustainable, either by behavior within the monopolist's control or by exogenous factors. The most solid result to date indicates that under a broad range of assumptions, a firm can render its natural monopoly sustainable by engaging in a form of elasticity-based tariffing known as Ramsey pricing,[11] which is a refined form of price discrimination. In words, Ramsey pricing prescribes higher prices for user groups with relatively price-insensitive demand schedules and lower prices for user groups that are relatively more sensitive to price.

If an incumbent firm is willing (and is allowed) to sustain its natural monopoly by appropriate behavior, including an effective tariff structure, there would appear to be little need for it to be protected against entry by statute or

11 The seminal reference is that of Ramsey (1927). The reinterpretations by Baumol and Bradford (1970) and by Boiteux (1971) are largely responsible for recent interest in inverse-elasticity pricing. Zajac (1978, Chapter 3) is an accessible conceptual, arithmetic and diagrammatic introduction to Ramsey pricing. Copeland and Severn (1985) oppose Ramsey and marginal-cost pricing and provide arguments for average-cost or unbundled pricing.

regulation. Any entrants (to the extent that they did not subsidize their losses) would be unable to establish market footholds, and their failure at the hands of market forces would confirm the existence of natural monopoly. The strongest case for regulatory intervention to protect a natural monopoly comes in those cases where such a monopoly is unsustainable. Society's overall welfare is certainly diminished if a given market basket of goods is produced at a higher cost by several firms instead of at a lower cost by a single firm, other considerations equal.

The case of INTELSAT is instructive. Many have argued that it should price its services more attractively and aggressively,[12] perhaps by revising its agreements to avoid the requirement that it price on the basis of average cost service by service. If it were to do this, it might be able to render its natural monopoly sustainable, to the extent that it is not sustainable already. Theory suggests that Ramsey pricing its services – which would admittedly present numerous political and practical difficulties – would remove any incentives that might induce latent rivals to enter and to compete in potentially remunerative submarkets. Thus, at least three regulatory policies are available to INTELSAT on the basis of the concepts of natural monopoly and sustainability already set forth. First, entry can be freely allowed, a case that seems appropriate for a sustainable natural monopoly. If INTELSAT's monopoly is not sustainable, two other policies suggest themselves. First, INTELSAT could be allowed or required to make its monopoly sustainable by using appropriate pricing strategies, although identifying the exact policies to be applied is a poorly understood process. Second, INTELSAT could be allowed to retain its present, more politically acceptable average-cost pricing regime and simply be protected by a prohibition of entry altogether.

Finally, this analysis has assumed that INTELSAT is indeed a natural monopoly. If entry is allowed without knowledge of whether the incumbent enjoys natural monopoly, an additional problem arises. Suppose that one or more rivals succeed in profitable entry against the incumbent. This could have one of two implications: first, that the incumbent does not enjoy natural monopoly (as suggested for the case of pre-divestiture AT&T by the work of Evans and Heckman 1983a, 1984); or second, that natural monopoly exists but is unsustainable. Given Snow's (1987b) findings with regard to INTELSAT,[13] successful entry of private competing systems into various internation-

12 Snow (1977b) interprets aspects of INTELSAT's pricing structure as reflecting marginal cost or inverse-elasticity elements.

13 The methodology introduced by Evans and Heckman, (1983a, 1984) has been commented on by Sueyoshi and Anselmo (1986), who demonstrate that the admissible region for subadditivity measurement is smaller than that claimed by Evans and Heckman, given the constraints that

al satellite markets might indicate that INTELSAT exhibits unsustainable natural monopoly and should either adapt more flexible pricing policies to deter such competition or be protected from competition altogether. The latter policy, as recent experience has shown, is politically unrealistic.

c. *Economies of scope*[14]

In a multiproduct setting, there may exist certain synergies or complementarities of the production process, called economies of scope, that make it cheaper for a single firm to produce an entire line of outputs than for individual firms to specialize in the production of one or a few of those outputs. It will be seen that both economies of scope and economies of scale (the latter after appropriate generalization to a multiproduct framework) can contribute to the property of natural monopoly, although in different ways. The case of joint production, in which one output is essential to the production of another, is a familiar special case of economies of scope. An example of joint production is the output by a single producer of cotton and cottonseed oil, or of hides and meat. In telecommunications, economies of scope might mean, for example, that a predominant network operator could offer terminal equipment services more cheaply than could a separate firm specializing in such services. Likewise, a single firm might exploit cost complementarities that allow it to offer voice, television and data services more cheaply than three firms could provide them separately. If we denote the level of these three services by V, T and D, respectively, the relationship for pure economies of scope is then given by

$$C(V,T,D) < C(V,O,O) + C(O,T,O) + C(O,O,D).$$

INTELSAT's apparent natural monopoly, if indeed it has one, might be supported in part by economies of scope from various sources. These might exist, for example, in the provision of global service to the three ocean regions by a single firm; in the provision of different kinds of service (telephone, video, data) by a single firm; or in the provision of a given service under different requirements or circumstances to different user groups (size of earth stations, interruptibility of service, etc.). Private systems competing for INTELSAT's Atlantic traffic (including links between North and South America) can be considered as attempting to exploit their own economies of

the latter propose. More substantively, Charnes, Cooper and Sueyoshi (1985) contend that the approach is sensitive to the methodology applied for a given cost-function specification and data set. By using methods of operations research, they arrive at a conclusion opposite to that of Evans and Heckman while using the latter's data and functional form.

14 Baumol and Braunstein (1977) and Pulley and Braunstein (1984) discuss conceptual and empirical aspects of estimating economies of scope.

scale in individual product lines (*e.g.*, certain high-density traffic streams) to overcome INTELSAT's inherent natural monopoly advantage due in large part to its economies of scope in multioutput production. To the extent these rivals succeed, they expose INTELSAT's natural monopoly as unsustainable, at least given its present pricing and other strategic policies.

d. *Contestability of market structure*

Modern economic theory has idealized the model of perfect competition, in which markets have large numbers of small buyers and sellers, none of whom is influential enough to affect price. This abstraction is clearly inappropriate to most telecommunications markets, even with the increased amount of competition and deregulation during recent years. Virtually all industrialized and developing countries have large public or private telecommunications enterprises, generally statutory monopolies, that can and do influence price by virtue of their size. Often large users can exert this same kind of influence.

Yet the static monopoly model familiar in regulatory economics is also no longer an appropriate characterization of most telecommunications markets, which since the 1970s have been subjected to increasing competition by smaller specialized firms.[15] Some large enterprises, like AT&T, have been split into their component parts; others, such as the major domestic carriers in Japan (NTT) and the United Kingdom (British Telecom), have been privatized, subjected to competition and revitalized. During this period of time, the industrial organization paradigm of perfectly contestable markets has emerged from the same strain of the economics literature that produced the natural-monopoly and sustainability models described above.

Like perfect competition, perfect contestability of market structure is an abstraction probably not observable in empirical reality. It seeks to account for the ability of competition, whether latent or actual, to exert the same kind of discipline on producers in a partially competitive market that it does under perfect competition. Such discipline, coming from autonomous market forces, is often more efficient and effective than government regulation or administration. Contestability theory has clarified the conditions under which competition, or the threat of competition, can succeed in positively affecting the behavior of the small number of large firms usually wielding market power in telecommunications markets.

15 Extensive country-by-country surveys of telecommunications deregulation are given by Snow (1986c) and by Bruce, Cunard and Director (1986).

Latent competitors must pose a credible threat of entry in order to exert such pressure on the production and pricing decisions of large incumbents. For such a threat to exist, rival firms must be able to enter the market easily (and at low cost) and, if they wish to do so, leave it easily as well. Under the idealized polar case of perfect contestability, there are no costs of market entry or exit.[16] The major source of such costs are sunk costs, which are investment expenses that cannot be recouped or redirected to other uses after their initial commitment. Thus, the question of whether or to what extent a given market is contestable turns on whether the market or technology in question requires sizeable sunk costs as a condition for market entry.

Scholars such as Jonscher (1986) have noted that British Telecom, for example, became considerably more competitive in its pricing and responsiveness to user requirements after the Mercury consortium was licensed to compete with it, even though Mercury's actual incursions into BT's markets have been minimal to date. Likewise, many observers have attributed INTELSAT's more competitive and cost-based tariff policies during recent years to the threat of competition from private systems, a threat that has still not concretely materialized. Yet casual empiricism suggests that satellite markets such as that dominated by INTELSAT are far from perfectly contestable. In other words, there are considerable sunk costs in establishing a communications satellite system.

Certainly the highest entry costs are incurred in designing and launching an entirely separate system. Such costs would be sunk to the extent that satellites could not be redirected to serve other markets if the market initially entered proved unremunerative. As a matter of fact, satellites do generally possess this flexibility. For example, a satellite originally intended primarily for voice traffic could be reconfigured in orbit to specialize in video or data services. Likewise, satellites can be drifted from one ocean region to another, or from one longitudinal location to another in the same ocean region, to link users in markets different from the ones first targeted. In addition, much lower entry costs are incurred when entrants engage in service competition rather than facilities competition. Service competition has become common in many of the more competitive telecommunications markets; it involves resellers of capacity and value-added service providers, who have been allowed by regulatory authorities to purchase, resell and, in some cases, enhance the value of the capacity they obtain from their incumbent competitors. INTEL-

16 Baumol, Panzar and Willig (1986) discuss the more empirically realistic case in which entry costs are small but positive.

SAT, for example, has declared itself potentially amenable to service competition as an alternative to full-scale facilities competition. If one assumes unsustainable natural-monopoly properties on INTELSAT's part, service competition becomes an attractive public-policy alternative.

For whatever reasons, however, competition with INTELSAT has been crystallizing along the lines of facilities, with the strong support of the U.S. government. In any event, the contestable market hypothesis is at least partially applicable to the case of INTELSAT; this is because the organization's marketing behavior appears to have been positively affected by the threat (if not yet the reality) of competition. This threat, in turn, has been rendered credible by the presumed ability of potential rivals to avoid sunk costs by redirecting their satellites to other services, regions and customers should they fail to establish a significant foothold against INTELSAT's primary users and traffic. Indeed, the U.S. government has not allowed private U.S. firms to compete against INTELSAT's switched traffic, and such firms have contended from the beginning that they would be competing primarily for traffic streams that INTELSAT was unable or unwilling to attract.

Chapter II: Outline of INTELSAT's Agreements and Institutional Structure

A. *Interim Agreements*

As satellite communications became commercially viable in the early 1960s, consultations were held among the world's industrialized countries regarding the most appropriate institutional arrangements for exploiting the new technology. A number of factors dominated these discussions. First, attention was focused almost exclusively on international traffic, most of which was transoceanic as well. Satellites were correctly perceived to have the greatest cost advantage over competing terrestrial modes (primarily submarine cables and high-frequency radio) when the greatest distance are involved, given the insensitivity of satellites to distance or transmission path.[1] Second, satellite technology was still sufficiently limited in both quality and quantity that the desirability of a single supplier was taken for granted. Third, the habits and instincts of the world's telecommunications officials, as well as the institutional and political realities then prevailing, dictated a governmental or quasi-governmental provision of international public satellite service. Fourth, the dominance of the United States in both satellite and launcher technology was a fact of life realistically accepted even by countries wishing to challenge that dominance at some later time. Fifth, this dominant U.S. role was seen, at least by Western European countries, as a passing phenomenon, one they did not wish to incorporate as a permanent feature in any institutional arrangements. Thus, INTELSAT's initial statutes were considered to be temporary or »interim« agreements, and their replacement by »permanent« or »definitive« treaties was carefully provided for. The negotiation of the interim agreements took place against the backdrop of initial successes in transatlantic communications using experimental or pre-commercial spacecraft. Particularly dramatic to the public at large were live international television broadcasts, which required too much bandwidth to be accommodated by the submarine cables then in use.

After a preparatory meeting in London in early 1964 a series of three multilateral agreements were opened for signature on August 20, 1964 in

1 Costs and other economic properties of communications satellites are treated at length by Snow (1987a).

Washington, D.C., where the new organization was to be located. These included an intergovernmental agreement (U.S. Department of State 1964, pp. 1–16) among the countries involved, known as parties, dealing with diplomatic and organizational matters; a »Special Agreement« (U.S. Department of State 1964, pp. 41– 54) among telecommunications operating entities designated by the governments involved, known as signatories; and an annex (U.S. Department of State 1964, reprint of January 1967) containing the draft supplementary agreement on arrangements for settling legal disputes that might arise in interpreting or applying the intergovernmental agreement or Special Agreement. Another annex to the Special Agreement (U.S. Department of State 1964, p. 74) contains a list of twenty countries and their designated entities as »prospective signatories to the Special Agreement.« The United States had created the Communications Satellite Corporation (Comsat) by an act of Congress in 1962 (U.S. Congress 1962) to serve as its signatory; France, the Netherlands, Spain and the Vatican designated their own governments; and the other countries designated their postal and telecommunications ministries (PTTs).

The separate annex to the Special Agreement also assigned each signatory an investment quota designed to approximate its presumed initial usage of the system. Comsat received a quota of 61.0 percent. The next highest were the United Kingdom (8.4 percent); France and West Germany (6.1 percent each); Canada (3.75 percent); and Australia (2.75 percent). Article XII of the intergovernmental agreement[2] provided that for the sixth-month period beginning August 20, 1964, the intergovernmental agreement would be open for signature by any government listed in the annex of the Special Agreement. In addition, any other country belonging to the International Telecommunication Union (ITU) could accede to the agreement subject to approval of a quota by members of the organization. Assignments of existing quotas were to be »reduced pro rata as necessary to accommodate additional signatories to the Special Agreement« (Article XII(c)).

The interim agreements did not bestow legal personality on INTELSAT as an international organization; this was accomplished only by the permanent agreements that were opened for signature in 1971 and entered into force in 1973. Technically, INTELSAT was a consortium during the interim regime. Indeed, it did not receive its present name until 1965.

2 Articles of the interim and permanent intergovernmental agreements are designated by Roman numerals, while articles of the (interim) Special and (permanent) Operating Agreements are designated by Arabic numerals.

The remainder of Section A outlines the major aspects of the interim agreements, particularly those with relevance for the negotiation of INTELSAT's permanent regime and for INTELSAT's challenges today.

1. *Cooperative financial structure*

Articles 3 and 9 of the Special Agreement contain the financial details of INTELSAT's early operation. The organization was established in a technical sense as a cooperative of owners and users,[3] a feature that has endured into the permanent regime and has contributed greatly to the businesslike and apolitical functioning of the organization up to the present.

In a cooperative structure of this kind, each signatory contributes to the »costs of the design, development, construction and establishment of the space segment« a percentage equal to its quota (Article 3). Article III of the intergovernmental agreement lays down, in addition, that INTELSAT's space segment »shall be owned in undivided shares by the signatories ... in proportion to their respective contributions to the costs ... of the space segment.« Furthermore, revenues from space-segment utilization charges allocable to »amortization and compensation for the use of capital« are to be credited to the signatories »in proportion to their respective quotas.« Thus, each signatory can be a net owner or a net user of INTELSAT, depending on whether its usage share of the system is less than or greater than its investment quota. Net owners, whose investment quota exceeds their usage, will have a financial interest in high utilization charges to maximize the return on their invested capital. Net users, whose usage exceeds their investment quota, will prefer low usage charges. Indeed, minor political problems arose in INTELSAT as these two groups emerged. Often the misalignment of quotas and usage shares resulted from overoptimistic estimates of earth station operational dates.[4]

It is obviously in the interest of a cooperative of owners and users to reallocate investment quotas periodically so that the financial incentives of these two roles are balanced for each member. This procedure is incorporated into Article 6(c) of the (permanent) Operating Agreement. Indeed, the level of prices charged for system usage is perfectly indifferent to an INTELSAT

3 A financial model of INTELSAT's cooperative structure is given by Snow (1976, pp. 147–49).

4 Article 6(d) of the Operating Agreement provides for the voluntary redistribution of investment shares among signatories.

member whose usage share and investment quota are exactly equal. As a practical matter, INTELSAT has historically aimed to earn a cumulative annual return of fourteen percent on its net invested capital so that net users can reimburse net owners for the use of such capital. The interim agreements do not fully address the question of aligning investment shares with usage. Article 8(c), to be sure, provides that in making allotments of satellite utilization, INTELSAT »shall give due consideration to the quotas of the signatories.«

2. *Comsat as manager of INTELSAT*

Passage of the Communications Satellite Act by the U.S. Congress in 1962 created the Communications Satellite Corporation and designated it as the U.S. entity in any international satellite venture. In addition, Article VIII of the intergovernmental agreement names Comsat as the »manager in the design, development, construction, establishment, operation and maintenance of the space segment.« Its duties as manager are spelled out in various articles of the Special Agreement. Article 9(c), for example, requires that Comsat collect revenues from INTELSAT users. Article 10 sets forth guidelines for Comsat's dealings with contractors and with INTELSAT's governing body. Article 12 charges Comsat, among other duties, to »recommend ... the type or types of space segment to be established«; to »operate and maintain the space segment«; and to prepare and submit »annual programs and budgets.«

As a practical matter, Comsat's dual identity as U.S. signatory and manager of INTELSAT provided the possibility, if not the reality, of various conflicts of interest.[5] Before the definitive agreements created a secretariat for INTELSAT, most INTELSAT staff members were provided in the form of Comsat personnel on loan. Given Comsat's head start in acquiring experience in satellite communications and the brief period of time envisaged for the interim regime, the selection of Comsat as manager probably solved more problems than it created. The permanent agreements, to be outlined in Section B below, provided for the gradual phasing out of Comsat's managerial role and the establishment of a permanent »executive organ« of full-time INTELSAT staff members.

5 Kinsley's (1976) treatment of Comsat's early years is detailed but unduly harsh. McDougall's (1985) prize-winning history of the U.S. space program contains a more balanced account.

3. *The Interim Satellite Communications Committee as INTELSAT's policy board*

Articles IV and V establish an Interim Satellite Communications Committee (ICSC) to function as INTELSAT's policy-making body. It is given responsibility for the »design, development, construction, establishment, maintenance and operation« of the space segment (Article IV(a)). It is comprised of representatives from each signatory with a quota of 1.5 percent or more and one from each group of smaller signatories agreeing to pool their quotas to attain a level of 1.5 percent or more (Article IV(b)). Article IV(c) establishes an advisory subcommittee on finance. Using the power of Article IV(d), enabling it to create »such other advisory subcommittees as it sees fit,« the ICSC formed subcommittees on technical and contracting matters. The latter met only episodically, while the financial and technical subcommittees were extremely busy throughout the interim regime dealing with their respective areas of expertise.

Article V(a) establishes voting power in the ICSC on the basis of quotas. Thus, INTELSAT's power structure was more like that of a commercial organization, in which ownership share determines control, than that of a democratically managed political entity functioning on the principle of »one man, one vote.« Quota voting persists in INTELSAT's permanent regime, although, as Section B will point out, two overarching policy boards provide one vote to each government and signatory irrespective of system usage or investment quota.

Article V(b) defines a quorum of ICSC members as consisting of representatives having total votes exceeding by 8.5 percent or more »the vote of the representative with the largest vote« (effectively Comsat, with an initial quota of 61.0 percent). Article V(c) exhorts the ICSC to »endeavor to act unanimously«; failing this, most decisions may be made on the basis of »a majority of the votes cast,« giving Comsat what amounted to a veto power. Subject matter requiring more than a simple majority of quota-based votes is spelled out in Article V(c) for fourteen policy categories. For decisions in these areas, the vote in favor must exceed that »of the vote of the representative with the largest vote« by 12.5 percent or more, effectively requiring a positive vote of at least four signatories, based on the initial quota allotments. Among these policy categories are included »choice of type or types of space segment«; »approval of budgets by major categories«; the »rate of charge per unit of space segment utilization«; »approval of placing of contracts«; »approval of quotas«; and »decisions relating to withdrawal« from INTELSAT.

The ICSC met approximately six times a year during the interim period, usually at meeting facilities leased from Comsat at its Washington, D.C. corporate headquarters.

4. *Contracting and procurement policy*

The criteria used by Comsat and the ICSC to recommend and to select contractors and subcontractors for space hardware were the greatest cause of dissention within INTELSAT during the interim period. For the INTELSAT I, II and III series, contracts were awarded overwhelmingly to U.S. firms. Western European and certain other INTELSAT members, however, orchestrated a carefully coordinated and ultimately successful campaign to win a greater share of contracting for the INTELSAT IV series. They did so in part by forming transnational consortia of aerospace firms to bid for INTELSAT subcontracts. INTELSAT's European members initially adopted a »fair share« approach to contracting, meaning that contracts should in general be awarded geographically on the basis of ownership shares. The United States and most developing members of INTELSAT resisted what they considered to be a subsidy of Western Europe's aerospace industry and argued instead for awarding contracts strictly on the basis of quality, price and timely delivery. In retrospect it appears that INTELSAT's European members, as a result of their forceful stance on the contracting question, gained the aerospace experience to make them competitive bidders in subsequent rounds. Western Europe, through the European Space Agency (ESA), now constructs and launches its own satellites; indeed Arianespace, ESA's commercial launch enterprise, has begun to launch satellites for INTELSAT. All prime contractors for INTELSAT series I through VI, however, have been U.S. firms, as Table 3 of Chapter I indicates.
Article X of the interim intergovernmental agreement, since it addresses the delicate balance between these conflicting philosophies of procurement policy in a transnational enterprise, is worth citing at some length:

> The Committee [ICSC] and the Corporation [Comsat] as manager shall be guided by the need to design, develop and procure the best equipment and services at the best price for the most efficient conduct and operation of the space segment. When proposals or tenders are determined to be comparable in terms of quality, c.i.f. price and timely performance, the Committee and the Corporation as manager shall also seek to ensure that contracts are so distributed that equipment is designed, developed and procured in the States whose Governments are Parties to this Agreement in approximate proportion to the respective quotas of their

corresponding signatories to the Special Agreement; provided that such design, development and procurement are not contrary to the joint interests of the Parties to this Agreement and the signatories to the Special Agreement.

In addition, Article 10 of the Special Agreement deals with Comsat's role in soliciting bids and in working with the ICSC for large contracts above specified amounts.

5. *Tariff policy*[6]

Article 9 of the Special Agreement provides what little guidance was available to the ICSC and Comsat regarding how INTELSAT's prices should be structured. Paragraph (a) sets forth the following:

> The Committee [ICSC] shall specify the unit of satellite utilization and from time to time shall establish the rate of charge per unit at a level which, as a general rule, shall be sufficient, on the basis of the estimated total use of the space segment, to cover amortization of the capital cost of the space segment, and adequate compensation for use of capital, and the estimated operating, maintenance and administration costs of the space segment.

In essence, these provisions imply the following charging principles. First, a single output is assumed (»the unit of satellite utilization«), which is conceptually one-half of a two-way, 4-kHz, voice-grade circuit. Other services, such as telex and television, had to be related to this basic unit. Second, charging was to be based on average cost. Third, the tariff framework was similar to rate-base public utility regulation in the United States and elsewhere,[7] which allows statutory monopolies to charge rates sufficient to capture an »adequate compensation« for use of its »capital,« i.e., rate base. Article 9(b) allows the ICSC to include in its costs those of Comsat's direct and indirect costs »which are allocable to its performance of services as manager,« as well as »appropriate compensation« to Comsat for such services, »as may be agreed upon between the Corporation and the Committee.«

6 See Snow (1976, pp. 47–96) for a detailed discussion of INTELSAT rate making through the mid-1970s.

7 See Snow and Jussawalla (1986, pp. 20–57) for annotated references and bibliographic essays regarding the literature on economic regulation and public utility economics.

6. *Access of earth stations to the space segment*

Articles 7 and 8 of the Special Agreement set forth conditions under which earth stations – owned in general by signatories but not part of INTELSAT's operating capital – may obtain permission to access INTELSAT's space segment. In general, signatories are responsible for submitting requests for access within their own countries. Article 8(a) enjoins each applicant for access to make »equitable and non-discriminatory arrangements for the use of the earth station by all signatories or duly authorized communications entities intended to be served by the earth station individually or jointly with other earth stations.«

7. *Provisions for the transition to permanent agreements*

Article IX(a) of the intergovernmental agreement requires the ICSC »in any case not later than 1st January 1969« to provide »recommendations concerning the definitive arrangements for an international global system which shall supersede the interim arrangements.« Among other things, the ICSC's report is to consider »whether the interim arrangements should be continued on a permanent basis or whether a permanent international organization with a General Conference and an international administrative and technical staff should be established.«

Article IX(b) establishes the aims of the definitive arrangements, including that all member-states also belong to the ITU; that investments to date be safeguarded; and that all parties be able to determine »general policy.«

The ICSC's report, according to Article IX(c), was to be considered at an international conference convened by the U.S. government within three months of its submission. Definitive arrangements were to be established by this conference »at the earliest practicable date, with a view to their entry into force by 1st January 1970.«

Article XV provides that the interim agreement remain in effect »until the entry into force of the definitive arrangements referred to in Article IX.« Likewise, Article 16 of the Special Agreement requires that it »continue in force for as long as the [intergovernmental] Agreement continues in force.«

B. *Permanent Agreements*

INTELSAT's permanent agreements, or »definitive arrangements« as they are referred to in the interim treaties, were painstakingly negotiated among representatives of INTELSAT countries during a two-year conference in Washington, D.C. They were opened for signature on August 20, 1971 and entered into force on February 12, 1973. As in the interim regime, two interlocking agreements were signed. The first, known as the intergovernmental agreement or the INTELSAT Agreement (U.S. Department of State 1971, pp. 1–81), is among governments, which are called parties to the agreement. It deals with INTELSAT as an international organization – its structure, financial principles and scope of activities. The second, known as the Operating Agreement (U.S. Department of State 1971, pp. 281–317), is signed, as during the interim period, by operating entities designated by member-governments. These entities are the signatories to the Operating Agreement, which deals primarily with the technical, operational and financial aspects of INTELSAT.

The rest of Section B outlines the principal features of the permanent agreements. Those differing substantially from the corresponding dispositions of the interim treaties will be given particular emphasis.

1. *Board of Governors as the executive body*

Article IX of the intergovernmental agreement establishes the Board of Governors as INTELSAT's principal executive and policy organ. A rather complex set of procedures is instituted to keep membership on the Board near twenty (paragraph (b)). In recent years, however, membership on the Board has been at or near twenty-seven. Members may be individual governors with a sufficiently high quota; governors representing INTELSAT members whose combined quotas are above the minimum amount; or governors representing five or more signatories from a given ITU region regardless of their combined quota (paragraph (a)). Paragraph (i) establishes a quorum for the Board as a majority of its members having two-thirds or more of the total (quota-based) voting representation of all signatories, or else »the total number constituting the Board of Governors minus three, regardless of the amount of voting participation they represent.« While enjoined to take decisions unanimously, the Board, failing this, is to vote on the basis of quotas. For all »substantive questions,« decisions are taken either by an affirmative vote of four or more governors comprising two-thirds or more of total voting participation, or by an affirmative vote »cast by at least the total number constituting the Board of Governors minus three, regardless of the amount of voting participation they represent.« »Procedural questions« are resolved by »an affirmative vote representing a simple majority of Governors present and voting, each having one vote« (paragraph (j)).

Article X sets forth the functions, responsibilities and powers of the Board of Governors. The Board is responsible »for the design, development, construction, establishment, operation and maintenance of the INTELSAT space segment and ... for carrying out any other activities which are undertaken by INTELSAT« (paragraph (a)). To indicate how these responsibilities are to be discharged, twenty-seven subparagraphs follow, outlining powers and functions specifically assigned to the Board. These include areas of procurement, financial policy, earth stations, charging policy, space segment access by non-members of INTELSAT, submission of reports and recommendations, conditions of employment in the executive organ, establishment of internal rules and many others.

2. *Assembly of Parties and Meeting of Signatories*

While the Board of Governors is vaguely familiar as a reconstitution of the ICSC under the interim regime, the permanent agreements create two completely new overarching policy bodies consisting of the entire population of parties and signatories, respectively, each taking decisions on the basis of one vote per member regardless of investment share.

The Assembly of Parties, created in Article VI, is further specified in Article VII. It is »composed of all the Parties [having signed the intergovernmental agreement] and shall be the principal organ of INTELSAT« (paragraph (a)). It »shall give consideration to those aspects of INTELSAT which are primarily of interest to the Parties as sovereign states« and »shall have the power to give consideration to general policy and long-term objectives of INTELSAT« (paragraph (b)). Paragraph (c) sets forth the Assembly's functions and powers, 14 of which are specifically enumerated. These include formulating its »views« or »recommendations« regarding »general policy and long-term objectives«; amendments to the permanent agreements; authorization of the use of specialized satellite facilities outside of INTELSAT; review of rules regarding non-discriminatory usage of the INTELSAT space segment; consideration and expression of views on various policy matters; taking of decisions »in connection with the withdrawal of a Party from INTELSAT«; consideration of complaints by parties; selection of legal experts for arbitration procedures; and appointment of a director general for INTELSAT. While paragraph (d) specifies a biannual meeting schedule for the Assembly, paragraph (e) lays down criteria and procedures for calling extraordinary meetings. Paragraph (f) prescribes a quorum as »representatives of a majority of the parties.« Further, »each Party shall have one vote.« Decisions on »matters of substance« are to be decided by a two-thirds affirmative vote, while »procedural matters« are decided by a »simple majority« of those present.

The Meeting of Signatories, also created by Article VI, is further specified in Article VIII. It consists of »all the Signatories« of the Operating Agreement (paragraph (a)). Paragraph (b) enumerates 12 »functions and powers« reserved for the Meeting of Signatories. These include expression of views and recommendations regarding reports, financial statements and proposed amendments; the »establishment of general rules« regarding earth stations, allotment of space-segment capacity and adjustment of charges for the INTELSAT space segment; decisions regarding withdrawal of signatories from INTELSAT; reports to the Assembly of Parties »concerning the implementation of general policies, the activities and the long-term program of INTELSAT«; and annual determinations »for the purpose of representation on the Board of Governors« in accordance with Article IX. The Meeting of Signatories is to meet »every calendar year« (paragraph (c)) in ordinary session or, at the request of the Board of Governors or with the support of one-third of the signatories, in extraordinary session (paragraph (d)). Voting rules and procedures are exactly analogous to those laid down for the Assembly of Parties (paragraph (e)).

3. *Replacement of Comsat as manager by a director general and permanent executive organ*

The permanent agreements provide for the phased replacement of Comsat as INTELSAT's manager by a »management services contractor,« which in turn is to be succeeded by an »executive organ« of professionals and staff working directly for INTELSAT. A transitional secretary general is ultimately to be replaced by a permanent director general, who will be assisted in his or her tasks by the executive organ of international civil servants. Article XV establishes the »privileges, exemptions and immunities« to be accorded to INTELSAT property, employees and representatives.

Article XII authorizes the Board of Governors to select a secretary general on a priority basis after entry into force of the agreements, who is to »be the legal representative of INTELSAT until the first Director General shall have assumed office« (paragraph (b)). The secretary general is appointed on the basis of »the highest standards of integrity, competency and efficiency« and »may be removed from office for cause by the Board of Governors« (paragraph (c)). The secretary general's functions are specified in Annex A to the intergovernmental agreement.[8]

For a period terminating six years from entry into force of the agreements, the Board of Governors is instructed to enter into a contract with Comsat as »management services contractor« for INTELSAT. The corporation is to perform »technical and operational management services« for INTELSAT as set forth in Annex B of the intergovernmental agreement. The contractor is to »be responsible directly to the Board of Governors until the assumption of office by the first Director General and thereafter through the Director General« (paragraph (e)). The secretary general is in effect an intermediary between the management-services contractor and the Board of Governors.

Paragraph (f) instructs the Board of Governors, within a year of entry into force of the permanent agreements, to commence a study »designed to provide the information necessary for the determination of the most efficient and effective permanent management arrangements« consistent with provisions of Article XI dealing with such arrangements and with the director general's responsibilities. No later than four years after entry into force, the Board of Governors is instructed by paragraph (g) to submit a report to the Assembly of Parties including the study just referred to, as well as »recommendations ... for the organizational structure of the executive organ.« By five years after entry into force of the agreements the Assembly of Parties, on the basis of this report and general provisions of the agreements, is to »adopt the organizational structure of the executive organ« consistent with Article XI, which deals with the director general and the permanent structure of INTELSAT (paragraph (h)).

The director general is to assume office »one year before the end of the management services contract ... or by December 31, 1976, whichever is earlier.« He or she is to be appointed by the Board of Governors and be »responsible for all management services ... and for the supervision of the performance of the management services contractor.« He or she »shall take all necessary steps to ensure that the permanent arrangements are fully imple-

8 Mr. Santiago Astrain of Chile served as INTELSAT's only secretary general, from 1973 until 1976, and as its first director general, from 1976 until his retirement in 1983.

mented not later than the end of the sixth year after the date of entry into force of this Agreement« (paragraph (j)).
Article XI provides information regarding the director general. He or she is to head the executive organ, which »shall have its organizational structure implemented by not later than six years after the entry into force of this Agreement« (paragraph (a)). The director general is defined as »the chief executive and the legal representative of INTELSAT« who is »directly responsible to the Board of Governors for all management functions.« Appointment to that office is »by the Board of Governors, subject to confirmation by the Assembly of Parties.« The director general »may be removed from office for cause by the Board of Governors on its own authority.«[9] The »paramount consideration« in selection is »the necessity of ensuring the highest tandards of integrity, competency and efficiency« (paragraph (b)). Paragraph (c) spells out the director general's power to negotiate, execute, administer and contract out technical and operational functions to »the maximum extent practicable with due regard to cost and consistent with competence, effectiveness and efficiency« (paragraph (c)). Paragraph (d) requires that the Board of Governors designate »a senior officer of the executive organ« to serve as acting director general »whenever the Director General is absent or is unable to discharge his duties, or if the office of Director General should become vacant.« A successor is to be found »as expeditiously as possible.«

4. *Provisions regarding the use of separate satellite systems by INTELSAT members*

Article XIV of the intergovernmental agreement contains provisions that were among the most hotly contested of any negotiated and that are crucial to many of the current issues to be taken up in Chapter IV. Paragraphs (c), (d) and (e)

9 This authority was in fact invoked and exercised by the Board in order to remove Mr. Richard R. Colino of the United States, who had become INTELSAT's second director general on December 31, 1983, from his office as of December 4, 1986. The Board also removed Mr. Jose Alegrett of Venezuela from his position as deputy director general at the same time. For details surrounding the removal of Mr. Colino and Mr. Alegrett, see, e.g., Wall St. J. Nov. 28, 1986, p. 13, »INTELSAT Auditor's Report Gives Details of a Controversial Payment to Brokers«; Washington Post, Dec. 19, 1986, p. D9, »State Dept. Seeks Results of Internal INTELSAT Probe.« Mr. Colino was replaced by Mr. John Hampton of Australia as acting director general in accordance with Article XI(d). On March 13, 1987, Mr. Dean Burch, former chairman of the U.S. Federal Communications Commission, was elected by the Board of Governors as INTELSAT's new director general. The Assembly of Parties, meeting in April 1987, confirmed his appointment to that position.

deal with the establishment, acquisition or use by INTELSAT members of separate satellite systems providing, respectively, »domestic public telecommunications services,« »international public telecommunications services« and »specialized telecommunications services.« To employ separate facilities for domestic use, a member must beforehand

> consult the Board of Governors, which shall express, in the form of recommendations, its findings regarding the technical compatibility of such facilities and their operation with the use of the radio frequency spectrum and orbital space by the existing or planned INTELSAT space segment (paragraph (c)).

Paragraph (d) deals more severely with separate international systems by applying an economic criterion as well. INTELSAT members contemplating the use of such systems must beforehand

> furnish all relevant information to and shall consult with the Assembly of Parties, through the Board of Governors, to ensure technical compatibility of such facilities and their operation with the use of the radio frequency spectrum and orbital space by the existing or planned INTELSAT space segment and to avoid significant economic harm to the global system of INTELSAT.

After such consultation

> the assembly of Parties, taking into account the advice of the Board of Governors, shall express, in the form of recommendations, its findings regarding the considerations set out in this paragraph, and further regarding the assurance that the provision or utilization of such facilities shall not prejudice the establishment of direct telecommunication links through the INTELSAT space segment among all the participants (paragraph (d)).

Paragraph (e) is much less stringent with respect to the use of separate specialized systems. Although it also provides for notification of the Assembly of Parties, there is no requirement to avoid economic harm. The Assembly must merely transmit to the Board of Governors »its findings regarding the technical compatibility of such facilities« and their operation with INTELSAT's use of the spectrum and orbit. Paragraph (f) requires the Assembly and the Board to conclude the coordination and notification procedures of Article XIV within six months after they begin. Paragraph (g) exempts separate systems used »solely for national security purposes« from the requirements of Article XIV.

5. *Tariff policy and procedures*

Article V of the intergovernmental agreement prohibits price discrimination among users of the same service: »The rates of space segment utilization charge for each type of utilization shall be the same for all applicants for the space segment capacity for that type of utilization« (paragraph (d)).
Article 8 of the Operating Agreement provides more specific guidelines for tariffing. Paragraph (a) requires that INTELSAT space segment utilization charges

> have the objective of covering the operating, maintenance and administrative costs of INTELSAT, the provision of such operating funds as the Board of Governors may determine to be necessary, the amortization of investment made by Signatories in INTELSAT and compensation for use of the capital of Signatories.

Paragraph (c) charges the Board of Governors, in determining the »rate of compensation for use of capital,« to »include an allowance for the risks associated with investment in INTELSAT« and, having done so, to »fix the rate as close as possible to the cost of money in world markets.« Throughout INTELSAT's operating history, the target cumulative rate of return on investment capital has been set at fourteen percent, although, as Table 2 of Chapter 1 indicates, the actual cumulative rate has been closer to fifteen or even sixteen percent in recent years.
In summary, Article V of the intergovernmental agreement and Article 8 of the Operating Agreement appear to constrain INTELSAT to non-discriminatory and average-cost tariffs, respectively. Strict adherence to non-discrimination would prevent INTELSAT from using any kind of elasticity-based pricing (some of which exhibits notable »second best« welfare properties in monopoly situations).[10] The average-cost pricing injunction of Article 8, however, is perhaps a less serious constraint on INTELSAT's pricing flexibility, since it refers only to overall prices and costs rather than to individual services and their associated costs. INTELSAT's pricing regime as it emerges from the permanent agreements is therefore once again, as in the interim period, consistent with the rate-base pricing principles used by regulated public utilities in the United States and elsewhere: non-discriminatory prices set at a level that covers operating and investment costs, depreciation and a »fair« or »market« return on net capital investment (the rate base).

10 See footnotes 11 and 12 of Chapter I.

6. *Settlement of disputes by binding arbitration*

Article XVIII of the intergovernmental agreement and Article 20 of the Operating Agreement describe categories of »legal disputes« that may or must be submitted to a procedure of binding arbitration set forth in detail in Annex C to the intergovernmental agreement. These disputes include those between or among INTELSAT, the parties and signatories to its agreements, and former parties and signatories. Disputes may involve interpretations of INTELSAT's agreements or of other agreements concluded separately.

Annex C contains the rules and machinery for effectuating the binding arbitration. Members of the Assembly of Parties may submit the names of no more than two »legal experts« (Annex C, Article 3(a)) able to serve on a »panel from which presidents of tribunals shall be selected.« The panel consists of eleven persons chosen by the Assembly from among those nominated. The panel, barring vacancies, ordinarily serves for the two-year interval between Assembly meetings (Article 3(b)). A quorum consists of nine panel members. Six or more panel members may choose a chairman by secret ballot (Article 3(c)). The Assembly is to constitute the panel so as »to reflect an adequate geographical representation, as well as the principal legal systems as they are represented among the Parties« (Article 3(e)).

Article 4 provides that a »petitioner« shall furnish »each respondent and the executive organ« with documentation of the subject matter of the dispute, justification of its appropriateness for arbitration under the agreements and a statement as to why consensus has not yet been reached by »negotiation or other means short of arbitration.« In addition, the petitioner provides »the name of [a] person desginated ... to serve as a member of the tribunal.«

Article 5 requires that within sixty days of receiving such documentation, »the side of the respondents shall designate an individual to serve as a member of the tribunal.« In addition, the respondents may »jointly or individually« reply to such documentation and present »counter-claims« if they desire. Failing the designation of a second tribunal member by the respondents, the chairman of the panel of tribunal candidates may designate a panel member other than himself or herself as a member of the panel. Finally, these two tribunal members are then to agree on a third tribunal member, chosen from among the panel members, to serve as president of the tribunal. Failing agreement on a tribunal president, the panel chairman designates a member of the panel other than himself or herself to serve in that capacity.

Articles 6 through 12 of Annex C contain rules of evidence and procedure to be followed by the three tribunal members, the petitioner and the respondents. Article 13 enjoins the tribunal to reach its decision based on »[the intergovernmental] Agreement and the Operating Agreement« and on »generally accepted principles of law.« If they so desire, the disputants »may reach an agreement during the proceeding« that »shall be recorded in the form of a decision of the tribunal given by consent of the disputants« (Article 7(g)). Otherwise, the tribunal reaches a decision in the absence of such an agreement. In either event, »[t]he decision of this tribunal, including any reached by agreement of the disputants..., shall be binding on all the disputants and shall be carried out by them in good faith« (Article 12(b)). If there is a »dispute as to the meaning or scope of its decision,« the tribunal »shall construe it at the request of any disputant« (Article 12(c)).

7. *Procurement and contracting*

Article XI(c) of the intergovernmental agreement mandates the following responsibility to the director general:

> The Director General, on behalf of INTELSAT, shall contract out, to one or more competent entities, technical and operational functions to the maximum extent practicable with due regard to cost and consistent with competence, effectiveness and efficiency. Such entities may be of various nationalities or may be an international corporation owned and controlled by INTELSAT. Such contracts shall be negotiated, executed and administered by the Director General.

Article XIII(a) sets forth important principles regarding the criteria under which alternative bidders are to be evaluated:

> [P]rocurement of goods and services required by INTELSAT shall be effected by the award of contracts, based on responses to open international invitations to tender, to bidders offering the best combination of quality, price and the most favorable delivery time.

Paragraph (b) provides that if such a combination is offered by more than one bidder, »the contract shall be awarded so as to stimulate, in the interests of INTELSAT, world-wide competition.« Thus, INTELSAT was able to avoid requirements for »fair share« or quota-based contracting, despite the pressure of certain European and other countries during the negotiation of the permanent agreements.

Article 16 of the Operating Agreement contains more specific procurement guidelines. Paragraph (b) requires approval of the Board of Governors for requests for proposals, invitations to tender and the award of any contract exceeding or expected to exceed $500,000. Paragraph (c) describes circumstances under which the board of Governors may waive its requirement to procure goods and services on the basis of »open international invitations to tender.« These include contracts for less than $50,000 (or a higher value, if so designated by the Meeting of Signatories at the request of the Board); a situation »where procurement is required urgently to meet an emergency situation«; circumstances under which »the requirement is of a predominantly administrative nature best suited to local procurement«; and cases in which

> there is only one source of supply ... to meet the requirements of INTELSAT or where the sources of supply are so severely restricted in number that it would be neither feasible nor in the best interest of INTELSAT to incur the expenditure and time involved in open international tender.

Article 5 of the Operating Agreement affects contracting procedures by designating a »capital ceiling« limiting the »sum of the net capital contributions of the Signatories and of the outstanding contractual capital commitments of INTELSAT« to an initial amount of $500,000,000. The Meeting of Signatories may, upon the recommendation of the Board of Governors, increase the ceiling. In addition, the Board itself may increase the capital ceiling by ten percent above its current level. The capital ceiling has been increased a number of times to allow for expansion of the INTELSAT system, which, as Table 2 of Chapter I indicates, had a net capital value of over $1,600 million at the end of 1985.

8. *Amendment of the agreements*

Article XVII of the intergovernmental agreement provides a mechanism for amending that document. Draft amendments may be proposed by any party to the agreement. The Assembly of Parties then considers the proposed amendments and may also »take decisions on any [other] amendment directly consequential to a proposed or modified amendment.« If approved by the Assembly of Parties, an amendment enters into force with the ratification of either two-thirds of INTELSAT's member-states having at least two-thirds of total investment shares, or eighty-five percent of INTELSAT's member-states regardless of their combined investment share. An amendment may not enter into force less than eight months or more than eighteen months after its approval by the Assembly.

Almost exactly parallel provisions for amendments to the Operating Agreement are contained in Article 22 of that document. »Any Signatory, the Assembly of Parties or the Board of Governors« may propose amendments. The Meeting of Signatories considers such amendment proposals and others »directly consequential« thereto. Following approval by the Meeting, an amendment enters into force upon approval by either two-thirds of the signatories having at least two-thirds of total investment shares, or 85 percent of signatories regardless of their total investment share. An amendment may not enter into force more than 18 months after its approval by the Meeting of Signatories.

C. *INTELSAT Membership Criteria and Conflicts*

Certain provisions of the permanent agreements are important in establishing and regulating the relationship between INTELSAT as an international organization and its constituent »members,« whether these are the states that signed the intergovernmental agreement or their designated signatories that signed the Operating Agreement. In addition, other provisions have direct or indirect implications for INTELSAT's relationships with other international organizations, with bidders and contractors and with user groups. Section C reviews the most significant of these relationships as they appear in the agreements.

1. *Withdrawal, expulsion and suspension*

Article XVI of the intergovernmental agreement is the basic document regarding sanctions against parties or signatories, as well as voluntary withdrawal from INTELSAT.

Paragraph (a) allows voluntary withdrawal of any party or signatory from INTELSAT. This action becomes effective three months after notification of such a withdrawal, or on the »date of the next determination of investment shares.«

Involuntary withdrawal of parties or signatories, which in essence is expulsion from INTELSAT, is covered in paragraph (b). In the case of parties, it provides that the Assembly of Parties may decide »that the party be deemed to have withdrawn from INTELSAT« if it finds that »failure to comply« with its obligations under the agreement »has in fact occurred.«

There are, by contrast, two separate categories of sanctions against signatories. One concerns non-payment of the capital contributions to INTELSAT required under Article 4(a) of the Operating Agreement. When such payments become three months in arrears, paragraph (c) of Article XVI provides that »the rights of the Signatory under this Agreement and the Operating Agreement shall be automatically suspended.« If another three months elapse after suspension without payment from the signatory, the Board of Governors »may recommend to the Meeting of Signatories that the Signatory be deemed to have withdrawn from INTELSAT.« The Meeting of Signatories may make such a decision and, from the date of that decision, »this [intergovernmental] Agreement and the Operating Agreement shall cease to be in force for the Signatory.« According to paragraph (e), there are three options for the party of a signatory having withdrawn or deemed to have withdrawn: the party may »assume the capacity of a signatory«; it may »designate a new Signatory as of the date of ... withdrawal«; or the party itself may withdraw from INTELSAT.

For signatories failing to comply with agreement provisions other than those regarding payment of capital contributions, paragraph (b) provides for the Board to recommend involuntary withdrawal to the Meeting of Signatories. The Meeting may then deem the offending signatory to have withdrawn, such withdrawal becoming effective immediately.

The remainder of Article XVI, as well as Article 21 of the Operating Agreement in its entirety, applies to financial and procedural arrangements during periods of suspension and following withdrawal.

A number of INTELSAT signatories have been had their rights under the agreements suspended for non-payment of capital contributions, including Chad, Honduras, Lebanon, Paraguay, Sudan, Tunisia, Uganda and Zaire. None, however, has been expelled (deemed to have withdrawn).

There have been only two voluntary withdrawals from INTELSAT. The Republic of China (Taiwan) withdrew in 1976 after it became clear that the People's Republic of China intended to accede to the INTELSAT agreements (see MS-6-10E, April 1977, p. 1).[11] The Republic of China received a settlement from INTELSAT according to the provisions of Article 21 of the Operating Agreement, and the investment shares of the remaining signatories were adjusted according to Article XVI of the intergovernmental agreement and Article 6 of the Operating Agreement. In 1977, the People's Republic of China acceded to the INTELSAT agreements. The second voluntary with-

11 Hereafter, INTELSAT documents will be referred to by number and date directly in the text. They are not included in the bibliography at the end of this study.

drawal was that of the Republic of Cape Verde in 1983, not long after its accession to INTELSAT. It was similarly reimbursed according to the applicable provisions of the agreements (see MS-14-10E (Rev. 1), 4 April 1984, p. 1).

2. *Representation in INTELSAT of non-parties and non-signatories*

During the interim period INTELSAT was technically a consortium. Under Article IV of the intergovernmental agreement, however, it received legal personality, including the ability to conclude agreements with states or international organizations, to contract, to acquire and dispose of property and to be a party to legal proceedings. Thus, INTELSAT is endowed with the legal power to interact with national governments, international organizations, private firms and courts of law, among other entities.

It is primarily through its director general, whose functions are spelled out in Article XI of the intergovernmental agreement, that INTELSAT deals with external individuals, organizations and other entities. In addition, at least two other international organizations are referred to in the INTELSAT treaties. The United Nations is mentioned in the Preamble of the intergovernmental agreement, where U.N. Resolution 1721 (XVI), urging the availability of satellites to the nations of the world on a global and non-discriminatory basis, is cited. The 1967 Outer Space Treaty, negotiated under U.N. auspices, is also mentioned there. In addition, the ITU is named in several places. For example, Article XIX(a) of the intergovernmental agreement requires acceding countries to be members of the ITU or to be parties to the interim agreements, which also required ITU membership. Article XVI(n), covering withdrawal, provides that »no Party or its designated Signatory shall be required to withdraw from INTELSAT as a direct result of any change of status of the Party with regard to the International Telecommunication Union.« Finally, Article 13 of the Operating Agreement exhorts INTELSAT to observe »the relevant regulations of the International Telecommunication Union« and to »give due consideration« to the applicable recommendations and procedures of the ITU's specialized committees in »the design, development, construction and establishment of the INTELSAT space segment and in the procedures established for regulating the operation of the INTELSAT space segment and of the earth stations.«

This being said, it is a fact that INTELSAT is not directly accessible to all groups and individuals whom it affects and who have an interest in it. For example, scholars and researchers technically have not had direct access to

INTELSAT records and documents, even though they are citizens of countries that belong to INTELSAT and provide its investment funds through tax revenues. Such information has been available only through those individual parties and signatories that chose to furnish it.[12] Aside from the PTTs that comprise most of the signatory organizations, user groups have also had no direct route of access. Non-PTT user interests include multinational corporations, trade organizations and consumer groups. The concerns of such entities, at least to the extent that they assume the form of documents brought to the attention of one or more of INTELSAT's deliberative bodies, must be channeled through the »organs« of INTELSAT set forth in Article VI of the intergovernmental agreement. The executive organ and the director general, for example, act as buffers between bidders and contractors wishing to provide goods and services to INTELSAT and the organs empowered to approve contracts, including in particular the Board of Governors and the Meeting of Signatories.

12 INTELSAT's policy regarding access to documents was formalized in 1984. The Board of Governors (MS-15-10E, April 2, 1985, pp. 8–9) noted that »INTELSAT documents have generally been considered ... relatively confidential in nature« in the past, and that »confidentiality has been maintained without the adoption of formal procedures such as those introduced by most other international organizations.« Noting that »procedures adopted in certain INTELSAT member countries« had resulted in »more ready access by the general public to INTELSAT documents,« the Board, at its September 1984 meeting, decided to adopt a classification of »INTELSAT Restricted« for documents »which would be undesirable to distribute outside Party or Signatory organizations because of the confidential or proprietary nature of the information contained therein.« Three criteria for such a classification were adopted:

> (i) information related to personnel matters; (ii) information which, if disclosed, could be used against INTELSAT's best interests (e.g., information concerning contract negotiations, market analysis, lists of customers or users, longer-term plans, or implementation of new services); (iii) information of a strategic, financial or operational nature (e.g., future charging policy, future network needs, etc.) (MS-15-10E, April 2, 1985, p. 8).

Documents not classified as »INTELSAT Restricted« by the director general according to one or more of these criteria are to be »unrestricted in their distribution.« The director general is to remove the »Restricted« classification »as soon as the sensitivity of the information ... has diminished to an adequate degree.«

It seems clear that the increased threat of competition from private systems and from fiber-optic cables has persuaded INTELSAT to take a more corporate, proprietary attitude toward certain kinds of strategic information.

Chapter III: INTELSAT's Policies and Practices

A. *Tariff Policy*

The discussion of tariff policy as an instrument of INTELSAT's overall business strategy requires two different perspectives. Subsection 1 deals with the pricing principles laid down in the permanent agreements and their relationship to the well-articulated literature on the theory and policy of public utility pricing.[1] Subsection 2 examines parts of INTELSAT's voluminous tariff manual in an effort to identify the extent to which the prices for its many individual services conform to the INTELSAT agreements, to welfare-maximizing criteria of public utility pricing, or perhaps to both.

1. *The permanent agreements in the light of welfare-maximizing pricing principles*

As Section B(5) of Chapter II noted, there are two provisions of the permanent agreements particularly relevant to establishing the principles of INTELSAT's tariff policy. Article V of the intergovernmental agreement prohibits price discrimination, while Article 8 of the Operating Agreement appears to require that INTELSAT's prices be based on average cost, at least in an aggregate sense. There is no mention that average-cost pricing need to apply service by service, however.

Average-cost pricing has three important properties that have endeared it to generations of legislators, administrators, accountants and regulators: it is easy to understand, it is simple to implement and it assures that revenues exactly cover expenses.[2] Under competitive conditions, of course, market forces automatically drive price (which the large number of buyers and sellers must confront as an exogenous influence rather than as a policy variable) to the level of average cost without the need for regulatory intervention. Average-cost pricing is the foundation of rate-base public utility regulation in the United States and elsewhere.

1 See Snow and Jussawalla (1986, pp. 58–66) for an annotated bibliography on pricing.

2 See Copeland and Severn (1985) for an iconoclastic defense of average-cost pricing and criticism of both marginal-cost and Ramsey pricing.

Economic theory, nonetheless, has established a strong case for the welfare-theoretic optimality of marginal-cost pricing under broad conditions.[3] If, the argument runs, price is under the control of a monopolist or regulator, it might be in society's overall interest to allow production up to the point at which price equals marginal cost. This is the point at which the amount that consumers are willing to pay for an additional (marginal) unit of output is just equal to the cost of producing that output. To stop short of that level of production would deprive both buyers and sellers of the welfare or benefits each group would derive from transactions beyond that output level up to the point at which marginal cost equals price. Beyond that point, of course, transactions would have a negative welfare effect, since the cost of producing each new unit of output (as measured by marginal cost) would be less than the value society ascribes to it (as measured by price).

Marginal-cost pricing principles were elaborated by economic thinkers well over a century ago but have not yet found widespread application in telecommunications or other public utility sectors of the economy. Notable attempts to introduce marginal-cost pricing principles have included experiments and occasional full-scale administrative implementation in the areas of peak-load and usage-sensitive pricing.[4]

Hesitation to implement marginal-cost pricing can be attributed to a number of factors. First, marginal costs, unlike total or average costs, are difficult to determine. Either quite arbitrary accounting principles must be used to allocate costs or the underlying cost function must be specified and econometrically estimated (and then differentiated). This problem is of course compounded when a multiproduct production setting, more relevant to the telecommunications sector, is employed. Second, and perhaps more important from a policy point of view, pricing at marginal cost will exactly recover total cost only if marginal cost happens to equal average cost. Marginal cost does equal average cost in the long-run equilibrium of the perfect-competition model. In sectors such as telecommunications with pronounced economies of scale and scope, however, marginal cost is generally still declining at the level of market output, which means that it is less than average cost and therefore that marginal-cost pricing will not produce enough revenue to cover total cost.

3 Kahn (1970, 1971) still provides the standard textbook defense of marginal-cost pricing principles and practice.

4 Mitchell (1978) presents a quantitative assessment of the costs and benefits of usage-sensitive pricing. Park and Mitchell (1986) arrive at a negative evaluation of usage-sensitive telephone pricing, finding that »contrary to the conventional wisdom, [it] is likely to be somewhat less efficient than traditional flat-rate pricing. If local measured service is desirable public policy, it must be justified on grounds other than economic efficiency« (p. v.).

This latter difficulty with the otherwise »optimal« features of marginal-cost pricing is not intractable. Beginning with Ramsey (1927) and his more modern reinterpretations by Boiteux (1971) and by Baumol and Bradford (1970), a number of authors have elaborated »second best« approaches to optimal pricing – generally in a monopoly setting where price is an endogenous policy variable rather than exogenously given. These methods, while based on marginal-cost principles, include certain auxiliary features to recover the revenue shortfall that such principles generally entail. In the strict formulation of Baumol and Bradford, user groups are differentiated according to their responsiveness to price changes (willingness to pay or, more formally, their price elasticity of demand). A higher price is charged to low-elasticity users, while a lower price is charged to high-elasticity users. In particular, each group must pay a percentage price markup from marginal cost that is inversely proportional to its price elasticity of demand, where the constant of proportionality is chosen so as to equate total revenue to total cost.

A more perjorative term for elasticity-based pricing is »price discrimination.« Such pricing, however, is common in all modern economies. Pricing structures that allow reductions for children and the elderly, for example, embody price discrimination, as do need-based educational grants, sliding-scale rate schedules for professional services and airline fares that distinguish between business travelers and tourists by imposing advance-purchase requirements.[5]

In telecommunications settings, elasticity-based pricing has met with criticisms of a theoretical and practical nature that have retarded its widespread implementation. Knieps, Müller and von Weizsäcker (1981), for example, note that Ramsey pricing in particular is politically divisive, since it charges different prices to different user groups for the same service and that it requires the often intractable estimation of price elasticities of demand for those groups. Yet Neumann, Schweizer and von Weizsäcker (1983) conclude that for the year 1978 alone, the Deutsche Bundespost could have provided a welfare gain of DM 1 billion to West German telephone users if it had applied Ramsey pricing by appropriately raising the price of local calls and reducing that of toll calls. Von Weizsäcker (1986) has observed that while elasticity-based pricing is often useful in static settings, a long-term dynamic environment generally argues for the use of cost-based tariffs.

INTELSAT's statutes acquire heightened policy interest when they are examined for impediments to different kinds of welfare-optimal charging policy. Thus, while Article 8, as noted, requires average-cost pricing, it does not appear to do so on a service-by-service basis but only in an overall sense.

5 Snow (1977b) interprets parts of INTELSAT's tariff as variants of price discrimination.

Indeed, a multiproduct setting is implied by its references to »charges« rather than to a single charge. Article 8 could in fact be construed as no more than an exhortation to recover total operating costs regardless of the pricing structure. Article V, however, presents more serious obstacles to welfare-efficient pricing. It is difficult to avoid the impression that it mandates non-discriminatory pricing, at least »for [the same] type of utilization.« Even here, though, there is leeway for imaginative »product differentiation« among what are in essence identical or equivalent services. In an economic sense, indeed, services are distinguished more by different patterns of user-group demand than they are by their innate physical properties. We turn now to an examination of INTELSAT's present charging policy as revealed in its tariff manual to examine the extent to which its pricing structure corresponds to the principles just enumerated.

2. *The structure of INTELSAT's tariffs*

INTELSAT's currently valid tariff manual (S-86-22, June 17, 1986) is a document of some 100 pages defining literally hundreds of different services and setting their prices. Yet a careful reading suggests numerous instances in which prices deviate significantly from the average-cost approach by incorporating principles of welfare-efficient rate formation, although no definite overall trend is perceptible.

By far the most fundamental and important aspect of the tariff manual is the charge for the lease of one unit of satellite utilization »for the establishment of one end of a two-way 4 kHz telephone circuit providing, as an objective, quality of service in accordance with appropriate CCITT/CCIR recommendations« (p. 1). Since 1981, as Table 2 of Chapter I indicates, the annual unit charge has been $390 per month, or $4,680 per year. Full-time telephone service has always constituted at least eighty percent of INTELSAT's overall revenues and traffic, although this share has continued to fall in recent years as video and data traffic and occasional service have become more important. The full-time telephone-circuit charge is an obvious example of average-cost pricing. Particularly in the early days of INTELSAT operation, it was derived essentially by dividing total forecast costs by total forecast traffic.

Other aspects of cost-based pricing also abound in INTELSAT's tariff. One pervasive source of this charging philosophy is found in the treatment of non-standard earth stations, which in general are smaller terminals requiring more powerful (and more expensive) amplification of the satellite signals they receive. »Rate adjustment factors« are applied to charging levels keyed for the standard A earth station. The tariff manual notes the following as a general principle:

> Satellite utilization through non-standard earth stations leased on a unit basis is charged for in multiples of appropriate rates applicable to similar use through standard earth stations. The appropriate rate adjustment factor to be applied will be determined on a case-by-case basis taking into account the particulars of the earth station and the actual services to be provided (p. 56).

The technical parameter used for such determinations is the gain-to-noise temperature ratio (G/K) of the earth station in question, which correlates with more tangible qualities, such as its diameter and cost.

For example, one-way preassigned single channel per carrier PCM/PSK circuits are charged at the rate of one-half unit of utilization with standard A stations but at one unit with standard B stations (p. 4). Likewise, standard B stations pay twice as much as standard A stations for equivalent one-way PSK data channels (p. 14). Another technical factor giving rise to cost-based charging differentials is the type of satellite beam.[6] A global beam radiates an entire third of the earth's surface (exclusive of polar regions) with relatively low power, while a more focused spot beam radiates a much smaller area with higher power. A hemispheric beam is somewhere in between. In addition, rates vary – by type of channel (simplex or half duplex), bandwidth, frequency range and types of modulation – in ways that generally reflect the pattern of cost variation attributable to the underlying technological parameters.

One important issue in charging philosophy arises from the multidestinational or broadcast mode that is inherent in the downlink of the satellite channel. On a cost basis, INTELSAT incurs the same expense whether a given signal is received by one, two or many more receiving stations. Thus, tariffs insensitive to the number of receiving stations can be judged as cost based, whereas those that increase measurably with the number of receiving stations are demand based. For example, voice-frequency telegraph bearer channels are charged only at the transmit end regardless of the number of receive ends involved (p. 2). For one-way FDM/FM service, however, each receiving station pays fifty percent of the charge for the transmit station, which is clearly a demand-based formula without a basis in underlying costs (p. 6). Likewise, occasional-use program channels are charged half the rate at each receiving station of that levied on the transmit end. Receiving stations, however, must also pay a

6 See Podraczky and Pelton (1984) for a general treatment of six generations of INTELSAT satellites and their beam characteristics.

per-occasion charge of $25, which reflects a marginal-cost adjustment to what is essentially a demand based tariff (p. 10). For preassigned single channel per carrier PSK circuits, the charge is again twice as much for the transmit end as it is for each of indefinitely many receiving stations. When a standard B station receives, however, the charge is double that for the standard A. This is an admixture of cost considerations in what is at bottom a demand-based tariff (p. 15).

The occasional rate for television is a curious case. The charge for a video half-channel has remained in the area of $8.00 per minute since the early 1970s,[7] while the underlying annual space segment utilization rate has fallen from $20,000 to $4,680. This means that whatever basis in underlying cost the occasional video rate might once have had no longer exists. Charging relatively more for video channels than for voice channels once again appears to be demand based, since undersea cables (at least prior to the current generation of fiber-optic cables about to be installed) do not have enough capacity to carry video channels. This lack of a competing mode, in turn, decreases the price elasticity of demand of video-channel users relative to that of voice-channel users, and it is the Ramsey pricing principle already referred to that mandates a higher price markup above marginal cost for lower-elasticity users and a lower markup for higher-elasticity users. An additional demand consideration in the video leasing charge is that each receiving station pays fifty percent of the charge of the transmit station (p. 19).

By contrast, tariff provisions for international television leased-channel service (pp. 22–25) yield only a slightly increasing revenue stream as more receive stations sign on. Perhaps this can be considered as a marginal-cost tariff reflecting the administrative costs of dealing with additional administrations. Users may consist of »one sending Signatory and any number of receiving Signatories,« or »any number of Signatories, all of whom shall have the right to send and receive.« According to the charging formula adopted, this service has the following price and revenue properties for two to five users:

Number of Users	Annual Charge per User	Total INTELSAT Revenue
2	$ 500,000	$ 1,000,000
3	375,000	1,125,000
4	312,500	1,250,000
5	275,000	1,375,000

7 See Snow (1976, pp. 51–54) for greater detail.

Thus, there is in effect an incremental revenue increase of $125,000 for each additional user.

One consequence of allowing prices to deviate from underlying costs, as in the case of demand-based pricing, is that incentives for cream skimming[8] and other forms of rate arbitrage are thereby created for users in a position to engage in resale of their circuits to third parties. In most cases of multidestination services where additional receivers must pay a fixed percentage of the transmit rate, INTELSAT's tariff policy prohibits such rate arbitrage. For example, one-way preassigned single channel per carrier PCM/PSK circuits are sold under the condition that »they will not be used to provide facilities similar to those of a network of point-to-point two-way circuits at a charge lower than that prescribed by INTELSAT for such circuits« (p. 5).

Similar proscriptions of such resale apply to multidestination one-way FDM/FM channels (p. 6) and to multidestination one-way preassigned single channel per carrier PSK circuits (p. 15). Likewise, the following provision regarding emergency double-hop transmissions during service restoration seems aimed at foreclosing the possibility of resale: »[T]he standard rate of charge ... will apply for each hop if the transmission is to be used and distributed at the intermediate point; otherwise, the standard rates applicable to a single-hop will apply« (p. 31-E).

INTELSAT has also differentiated its output into services reflecting various levels of quality and reliability, thereby allowing it to capture subtle patterns of demand variation that would be lost in a more monolithic tariffing structure. An excellent example is provided by the prices of various categories of non-preemptible leased international video services (pp. 29–30). Category A, provided over primary- and major-path satellites, is charged at a rate »twice that of the applicable preemptible service.« Category B, provided over other satellites, is charged at a rate »75% over and above the rate of charge applicable to the preemptible service«. Category C is like Category B except that the former »would not be subject to full restoration and normally would receive only partial restoration in case of a transponder or satellite failure.« Category C service is tariffed at »50% to 75% ... over and above the rate of charge applicable to the preemptible service,« depending on »the degree of protection provided.«

Finally, INTELSAT has captured patterns of demand variation in various »concessionary« or »promotional« rates that have been introduced from time

8 Kahn (1971, pp. 220–46) deals at length with institutional aspects of cream skimming.

to time, often in the interest of marketing unused or excess capacity. A notable example of this is the lease of whole, half and quarter transponders to signatories for strictly domestic purposes on an interruptible basis.[9] Originally, whole transponders were offered under these conditions for $1,000,000 per year, which was considerably below the then-prevailing price for an equivalent number of full-time leased circuits. As of June 1985, the price fell to its present level of $680,000 for a whole transponder in »spare capacity,« compared with a charge of 360 units ($1,684,800, at current rates) for a lease of »regular capacity« (pp. 32–33). Adjustments are made for fractions or multiples of a transponder and for the type of beam and satellite used. A »promotional lease rate« is also offered for the transition from domestic to international service using preemptible video services on satellites other than those used for primary- and major-path transmissions (pp. 26–26A). In the first year, international service is tariffed at a level of twenty-five percent of the difference between the domestic and international rate added to the domestic rate; in the second year, another twenty-five percent of the difference is added to the rate for the first year; and in the third year and beyond, the international rate applies. Clearly, INTELSAT costs are insensitive to whether a given amount of capacity is used for domestic or international transmission. In addition, it is evident that the users' price elasticity of demand will in general be lower for international than for domestic service, since Article XIV of the intergovernmental agreement makes it more difficult to establish separate international systems than separate domestic systems. Thus, the Ramsey pricing principle is at work here, since the low-elasticity users are being charged more than the high-elasticity users for what is in essence a technically equivalent service.

A final word is in order about the increasing diversity of the INTELSAT tariff and the swiftness and precision with which it is now responding to a broad variety of user needs. Paramount among recent innovations is the INTELSAT Business Service (IBS),[10] a »totally digital integrated service with no distinctions made between specific service applications.« Its primary object is for use with »networking.« The VISTA service,[11] by contrast, is for use in »rural and remote communities.« Designed for the generation of recently available, modestly priced and moderately sized earth stations, the VISTA service can transmit voice, telex, teletype or low-speed data for both domestic and

9 The origin of this policy in the early 1970s is discussed by Snow (1976, pp. 65–68).

10 INTELSAT Tariff Manual, S-86-22, June 17, 1986, pp. 41–48. See also Perras (1984, pp. 258–61) for more detail on IBS and other recently introduced services.

11 INTELSAT Tariff Manual, S-86-22, June 17, 1986, p. 49. See also Perras (1984, pp. 261–64) for more detail on VISTA service.

international users. Finally, INTELNET (pp. 50–54) is a new class of service for very small, inexpensive earth stations transmitting data. The terminals, which can be as small as 65 cm in diameter, are well suited for rural and remote areas from which weather information, emergency messages, news or administrative material must be transmitted. Interactive data is available via INTELNET II terminals, which can be as small as 2.5 meters in diameter.

In conclusion, there is every indication that INTELSAT has moved in the direction of more innovative and efficient pricing in recent years. This subsection has presented evidence that individual tariff offerings are based on marginal cost and on demand patterns as well as on average cost. In addition, the sheer volume and diversity of INTELSAT's tariffed services have increased considerably during recent years. It has been asserted that these developments have come too late, are not sufficiently comprehensive and have been reactions to the external pressure of actual and planned competing systems rather than independent initiatives on the part of INTELSAT's planners and constituents. We shall discuss the validity of such criticisms throughout Chapter IV.

B. *INTELSAT's Contribution to the Peaceful Use of Outer Space*[12]

INTELSAT is an organization that both by design and by its actions has contributed to the peaceful use of outer space and to more productive and efficient interaction among the nations of the world. Furthermore, these peace-promoting activities have not been confined to the realm of purely diplomatic exchanges of views or merely procedural cooperation. Since it began operation in 1965, INTELSAT has drawn more than 100 countries into a cooperative, commercial, operational enterprise that has invested a net amount of more than $1,000 million into the first (and to date only) profitable space technology.

INTELSAT's interim and permanent agreements are explicit, if brief, about its pacific role. The preambles of both intergovernmental agreements refer to United Nations Resolution 1721 (XVI) to the effect that »communication by means of satellites should be available to the nations of the world as soon as practicable on a global and non-discriminatory basis.«

12 See Snow (1986b) for a comprehensive treatment of INTELSAT's contributions to world peace.

In addition, the preamble of the permanent intergovernmental agreement mentions the Outer Space Treaty signed under United Nations auspices in 1967, according to Article I of which »outer space shall be used for the benefit and in the interests of all countries.« Additional language in the preamble of the permanent intergovernmental agreement is worth citing at length, since it appears in a document that is otherwise largely concerned with the organizational mechanism and financial arrangements of a commercial, profit-making enterprise:

> The States Parties to this Agreement ... [d]esiring to continue the development of this telecommunications satellite system with the aim of achieving a single global commercial telecommunications satellite system as part of an improved global telecommunications network which will provide expanded telecommunications to all areas of the world and which will contribute to world peace and understanding, [d]etermined, to this end, to provide, for the benefit of all mankind, the most efficient and economic facilities possible consistent with the best and most equitable use of the radio frequency spectrum and of orbital space, [b]elieving that satellite telecommunications should be organized in such a way as to permit all peoples to have access to the global satellite system ... [a]gree as follows...

At least two features of this declaration are noteworthy. First, reference is made to »the« and to »a single« global commercial telecommunications satellite system, rather than to many such systems.[13] Second, the juxtaposition of lofty goals of world peace and understanding with the presumably more mundane task of providing »efficient and economic« facilities reflects not only the determination of INTELSAT's members to harmonize a commercial and operational mission with the peaceful use of outer space but also their belief that the former can in fact contribute to the latter.

Evaluating INTELSAT's activities and accomplishments during the past two decades is a necessarily subjective exercise, but it is one that nonetheless can provide persuasive evidence that INTELSAT has on balance functioned in and promoted the cause of peaceful use of outer space. By extension, it has thereby contributed to peace and understanding in general among the nations of the world. A number of general frameworks within which INTELSAT's performance can be assessed will be suggested and evaluated in Section A of Chapter V. We shall now examine five aspects of INTELSAT's contributions to peace.

13 Snow (1985b), using Pacific rim and basin countries as examples, argues that the evolution of telecommunications systems from monopoly to deregulation is a natural and predictable one for technological, economic and institutional reasons.

The first focuses on developing countries. Although INTELSAT was criticized early on, with some justification, for being insufficiently sensitive to the needs of the Third World,[14] it has shown increasing concern for developing regions in recent years.[15] This concern has been manifested in three primary ways: more flexible operational and tariff policies regarding the use of smaller, less expensive earth terminals; a greater variety of low-cost services designed primarily with the needs of developing countries in mind (VISTA, INTELNET, Project SHARE); and of course provision in the permanent INTELSAT agreements for a pair of one-man, one-vote bodies (the Assembly of Parties and the Meeting of Signatories) that give a greater voice to INTELSAT's members, the majority of which are located in developing areas.

A second peace-related aspect of INTELSAT's structure and operation involves the cooperative financial principles along which it was organized in both the interim and permanent regimes. In effect, its organization as a cooperative of owner/investors and users has replaced competition with incentives for cooperation toward a common goal, which is the most efficient use of the space segment. It is perhaps this aspect of INTELSAT that has been most threatened by the imminent competition of separate systems. Such rivalry, according to allegations by INTELSAT and most of its members, will drain traffic away from the organization and thereby (given its presumed natural monopoly properties based on economies of scale and scope) make its circuits more expensive for remaining users. The validity of these claims will be examined at length in Section A of Chapter IV.

Third, INTELSAT's highly diverse membership has found itself in a quite specialized technical and operational setting in which the goal of an efficient and commercially successful global system has required not only cooperation in the sense of working together but also a certain de-emphasis of political factors as well. This phenomenon comports well with the theory of functionalism in international relations, which one of its exponents, Ernest B. Haas, has described as follows:

> Functionalists ... are interested in identifying those aspects of human needs and desires that exist and clamor for attention outside the realm of the political. They believe in the possibility of specifying technical and non-controversial aspects of governmental conduct, and of weaving an ever-spreading web of international institutional relationships on the basis of meeting such needs. They would concentrate on commonly experienced needs initially, expecting the circle of the non-controversial to expand at the expense of the political, as practical cooperation

14 *See, e.g.,* Hudson, *et al.* (1979, pp. 70–74).
15 *See, e.g.,* Pelton (1982).

became coterminous with the totality of interstate relations. At that point, a true community will have arisen (*quoted in* Pelton 1974, p. 11).

It bears mention that INTELSAT numbers among its members several countries that outside of the organization have been involved in local or regional conflicts or have otherwise been embroiled in international political controversy. These include Afghanistan, Angola, Iran, Iraq, Libya, Nicaragua, South Africa and Vietnam. The issue of international commercial satellite systems apart from INTELSAT has been controversial and upsetting to most INTELSAT members precisely because it is regarded as the intrusion of a politically charged issue into a stable, well functioning and above all apolitical environment.

INTELSAT's function as a role model for other international organizations is a fourth instance of its contributions to peace. In particular, INMARSAT[16] was consciously structured along the lines of INTELSAT and has functioned successfully since it began operation in the late 1970s. In addition, the Soviet Union is one of the major members of INMARSAT. One can speculate with cautious optimism about the example that INTELSAT can offer to future commercial, intergovernmental enterprises as well. These will involve inherently extraterritorial resources and technologies on the verge of commercial fruition. The exploitation of the Antarctic[17] and of the sea bed come to mind, although the failure of the Law of the Sea conferences is a warning that agreements in this area are not easy to come by.

Finally, having noted the Soviet Union's membership in INMARSAT, we must come to terms with the failure of INTELSAT to include as members the Soviet Union and many of its political allies in Eastern Europe and elsewhere. Of the countries even remotely fitting this description, only Afghanistan, Libya, Nicaragua, Vietnam and Yugoslavia are INTELSAT members. The Soviet Union and a number of countries with which it has close commercial and political ties belong to INTERSPUTNIK,[18] an international satellite organization with headquarters in Moscow. Operationally, INTERSPUTNIK is little more than an overseas extension of the Soviet Union's domestic satellite system. Since these two organizations serve what are essentially mutually exclusive streams of traffic, there is little conflict or rivalry (and

16 For complaints about competitive threats to INMARSAT, see Lundberg (1986).

17 Shapley (1985) gives a comprehensive account of the political and resource issues presently confronting the exploration and exploitation of the Antarctic.

18 Galloway (1972) gives an early comparison of INTERSPUTNIK and INTELSAT.

almost no contact) between them. Nonetheless, the inability to accommodate such traffic into a single, truly global, organization reflects world political realities of the 1960s that have changed little since that time.[19]

C. *INTELSAT's Use of Scarce Resources*[20]

Satellites, like terrestrial wireless technologies, must use the radiofrequency spectrum to provide point-to-point and broadcast services. Unlike those technologies, however, satellites make additional demands on another scarce resource: positions in the geostationary orbit. Section C explores the nature of INTELSAT's use of these resources; criteria for evaluating the effectiveness and efficiency of that use; and a possible regime for their more optimal utilization.

1. *Orbit and spectrum*

In addition to their obvious physical differences, these two scarce resources exhibit other contrasts as well. As Martinez (1985, pp. 149–61 *passim.*) notes, the spectrum is »universally accessible,« while the geostationary orbit is not. In other words, apart from regulatory complications, one can physically access the spectrum with ease and at low cost, while physically accessing the orbit is presently an exercise requiring on the order of $100,000,000, including the costs of a launcher, a spacecraft and associated hardware. In the language of the collective-goods model – to be discussed in greater detail in Section A of Chapter V – the spectrum is »indivisible and nonexcludable« and thereby a collective good, while the orbit, being both divisible and excludable, is not a collective good. The addition of a good such as the orbit to the existing storehouse of scarce communications resources has introduced significant new political tension into the existing international regulatory framework. This affects the ITU in particular, which has a well-articulated and smoothly functioning apparatus to accommodate conflicting claims on the spectrum, but

19 McDougall (1985) gives a comprehensive history of the continuing Soviet-American rivalry in space.

20 This section draws on a recent paper by Levin (1986), who has written extensively about theoretical and institutional issues in allocating the spectrum and orbit, as well as concrete procedures for doing so. *See also* Levin (1965, 1970, 1971, 1984, 1985).

whose members have not yet been able to agree on a comprehensive and permanent method to allocate orbital space.[21] Because access to the orbit is still effectively on a »first come, first served« basis, countries unable to reach the orbit themselves must either rely on INTELSAT to do so on their behalf or must contract out satellite construction and launch services to countries or firms that have the requisite experience and expertise. The latter solution is generally much more expensive, although various countries have paid the price in return for what they have judged to be greater benefits in terms of national prestige and security. For countries depending on INTELSAT for reasons of cost and the organization's perceived apolitical nature, the separate-systems issue is once again an upsetting and divisive one, involving as it does centrifugal and political currents that are seen as striking at the heart of the organization's purpose and mission.

2. *The evolution of orbit and spectrum allocations for satellite systems*

Although INTELSAT has blazed new trails as a commercial, operational international organization structured as a financial cooperative of owners and users, it cannot claim to have been similarly innovative or imaginative in its approach to spectrum and orbit usage. Indeed, by restricting its membership to states belonging to the ITU, it has implicitly adopted the ITU's long-standing agreements and machinery for allocating the spectrum. Naturally, the ITU has adapted its regulations and the functions of its specialized committees and periodic administrative conferences to the new and unique spectrum requirements of satellite communications. Nonetheless, these have been marginal adjustments within an existing framework rather than radical departures from past thought and practice.

In the case of orbital usage, by contrast, INTELSAT in 1965 confronted an unprecedented situation: it was exploiting a new resource for the first time, one that was not yet scarce. Consequently, no international agreements or organizations existed to regulate its use. During the interim regime, when the geostationary orbit was available in such abundant supply, little thought was given to its efficient management. The preamble of the interim intergovernmental agreement, for example, refers to »the best and most equitable use of the radio spectrum« but does not mention the geostationary orbit. By the early 1970s, when the permanent agreements had entered into force, use of

21 Martinez (1985, pp. 53–82) outlines in detail the current patchwork of regulations and procedures for securing orbital space.

the orbit by both INTELSAT and domestic and regional systems had progressed to an extent sufficient to sensitize the organization to the scarcity and consequent value of that resource. Indeed, the new intergovernmental agreement refers in its preamble to »the best and most equitable use of the radio frequency spectrum *and orbital space*« (emphasis supplied).

During recent years, there has been a gradual evolution to a more ordered environment for orbital planning and allocation. At the 1979 World Administrative Radio Conference (WARC-79) a number of countries forcefully urged changes in the ITU radio regulations in order to establish not only an abstract right to orbit and spectrum but also a guarantee that each country would be able to obtain its fair share of both those resources when they were needed (Withers and Weiss 1984, p. 305). To realize these objectives, the ITU scheduled two six-week sessions of a World Administrative Conference on the Use of Geostationary Orbit (Space WARC), one of which took place in 1985 and the second of which is scheduled for 1988.

Although the 1985 Space WARC began with a certain amount of polarization between industrialized and developing countries, the meeting registered the first real progress to date regarding the use of the geostationary orbit (*see generally Space WARC '85: A Look Back*, Satellite Communications, November 1985, pp. 67–70). Initial efforts by the industrialized countries thwarted attempts led by a bloc of LDCs to plan as many services in as many frequency bands as possible. As a compromise, discussion was limited to the fixed-satellite service; the number of bands was restricted to the expansion bands in the 4/6- and 11-12/14-GHz bands chosen at the 1979 WARC; and total new bandwidth was kept to 800 MHz. In addition, certain equatorial countries, notably Colombia and including Ecuador, Indonesia and Kenya, pressed claims to sovereignty over the portion of the geostationary orbit above their territorial boundaries.[22] The Western industrialized countries, along with the Soviet Union and its political allies, were successful in removing these claims from the agenda, using the argument that they were not within the ITU's competence, being perhaps more relevant for consideration by the United Nations.

A middle ground finally emerged that served as the basis of a compromise. Each ITU member was allotted one orbital position and some bandwidth as well. The bandwidth applied only to the expansion bands just mentioned. This decision successfully attenuated the differences and tensions between the LDCs at the conference, who had favored imposing a system of rigid a priori

22 Such claims were first formally asserted in the Bogota Declaration of 1976. See Gorove (1984) for a general discussion.

planning on orbital configurations, and the United States, which led other industrialized countries in favoring a more flexible regime. A paper circulated by the U.S. delegation assuaged anxieties of the LDC delegations by providing assurances that the orbital positions to be allocated would not be transferable or exchangeable.

At the same conference, INTELSAT, with the support of several national delegations, attempted to bring about a decision granting it preferred status in planning, an outcome that would have disadvantaged private international satellite systems in obtaining the spectrum and orbital arc necessary for sustained operation. Efforts by the U.S. delegation, however, succeeded in keeping the final acts of the conference noncommittal with respect to the relative status of multilateral, regional, domestic and private systems.

3. *Possible market-like approaches to allocating orbit and spectrum*

Rigid a priori planning characterizes the approach to orbit and spectrum allocation favored by most developing and many developed countries. This comports with generally accepted engineering, legal and political criteria. Engineers favor tightly packed, heavily used spectrum assignments. Jurists oppose any kind of regime that seems to accord permanent ownership rights to facilities in outer space in possible contravention of the 1967 Outer Space Treaty. Political considerations militate against any scheme that appears to imply a surrender of sovereign national entitlement to part of the orbit or spectrum.

Closer examination of the issues involved, however, suggests that application of economic principles to the previously intractable problem of orbit and spectrum allocation and use has much to recommend it. In essence, temporary entitlements to the use of orbit and associated spectrum, decided upon periodically by the ITU, could be traded in a loosely regulated »quasi-market« for such entitlements. Presumably, most of the buyers would be industrialized countries, and most of the sellers would be LDCs. Some semantic delicacies are involved; one would wish to speak of »permits,« »assignments« or perhaps »licenses« rather than »rights« being bought and sold. Nonetheless, as Levin argues:

> We must distinguish between the engineer's standard of full occupancy under a tightly packed plan (with which transferable assignments seem inconsistent), and the economist's standard of economic efficiency under a loosely packed plan. In the latter case, transferability is not only conceivable ... but may also lead to

> economically superior results. The task is simply to estimate and compare the economic losses in foregone outputs due to the vacancies left for flexibility, and the greater net value that results from transferring assignments from lower to higher valued users and uses (Levin 1986, p. 9; emphasis in original omitted).

Striking precedents exist for the creation of market-like regimes to assign orbit and spectrum allotments. Crutchfield (1961), for example, chronicles the emergence of quasi-markets in fishery licenses and assesses their effect on allocative efficiency and consumer welfare. Hahn and Noll (1983) describe the theoretical and empirical considerations leading to the implementation by the U.S. Environmental Protection Agency of a trading program in emission (pollution) permits. In addition, there is interest in the potential of market-like mechanisms to assign landing rights for passenger airlines at airports (Levin 1986, p. 4).

In terms of the communications industry, the purchase and sale of commercial radio and television stations in the United States and elsewhere are excellent examples of vibrant and smoothly functioning markets in entitlements that are ultimately based on spectrum assignments and broadcast licenses issued by regulatory bodies such as the Federal Communications Commission (FCC).[23] It should be noted that the regulator is not paid for the »right« to use the spectrum. It is the value conferred on the original licensee to use the spectrum for a finite period of time (until the next relicensing) that gives the station a market value over and above that of its physical plant and equipment. There is certainly no implication that the spectrum itself is being bought or sold, permanently or otherwise. Purchases of licensed radio and television stations in an open market are the mechanism whereby access to the spectrum is transferred from lower to higher valued uses over a period of time lasting until the license must be renewed.

To allay the fears of many LDC delegations at the 1985 Space WARC, the United States, as noted above, issued a surprising paper purporting to demonstrate that transferable spectrum and orbital assignments would seriously impair the existing allocation system (Levin 1986, p. 9). Nevertheless, market-like mechanisms for transferring spectrum and orbital allocations have arisen spontaneously in the international satellite industry itself. INTELSAT, for example, has leased transponder capacity for domestic use and at concessionary rates to a number of countries, most of them less developed ones, since the early 1970s. The lessees are free to configure and use this capacity as they wish for domestic applications.

23 See Levin (1980) for an economic treatment of licensing by the FCC of U.S. commercial television stations.

Levin (1986, p. 15) asks whether transactions in the opposite direction might not also be sanctioned. In addition to »transponder leasing or outright sale to LDCs for domestic purposes, with their implied leasing of associated orbit spectrum«, might it not be possible to »justify recognition of the reverse,« namely, »the subleasing of unused LDC assignments to [developed country] concessionnaires who seek more«? This kind of transaction already appears to have occurred in the case of the proposed joint venture between the government of Papua New Guinea and PACSTAR, a Pacific basin satellite system designed by Pacific Satellite Inc. (PSI), a subsidiary of the U.S. aerospace firm TRT. In this case, Papua New Guinea notified the ITU's International Frequency Registration Board (IFRB) on behalf of PSI that PSI would be using two of Papua New Guinea's ITU orbital assignments in return for free domestic circuits from the satellite operator (*see* Satellite News, March 11, 1985, pp. 1–2). In a comparable case, Telesat Canada agreed to tilt its ANIK-C spacecraft southward for use by the U.S. firm RCA-GTE for a price of $2 million per year (*see* Satellite Week, May 16, 1983, p. 3).
Certainly there are legal, technical and financial peculiarities of cases such as these that make it difficult to draw universally valid conclusions. The lesson for INTELSAT is nonetheless clear. Just as it has shown flexibility in its pricing policy for individual services, it can defend its own interests as a multinational, intergovernmental satellite system by adopting a judiciously flexible approach to the assignment and possible transfer of orbit and spectrum among ITU allottees. It might transfer some of its excess or unused capacity, for example, to individual INTELSAT members so that those wishing to do so could establish their own separate domestic systems. Conversely, it might negotiate with individual countries with a view to using their allotments of spectrum and orbit if additional capacity were needed for the global system. The financial, operational, technical and legal arrangements for such transactions will no doubt pose challenges, but this middle ground of allowing voluntary, welfare-increasing transactions among willing buyers and sellers in orbit and spectrum »quasi-markets« seems to promise more than either the polar case of rigid a priori planning or the other extreme of »first come, first served.«

Chapter IV: Major Issues Affecting Relationships Between INTELSAT and Its Member-Countries

Drawing on the historical, institutional, technical and economic background developed in Chapters I through III, Chapter IV presents and discusses in detail the most important current challenges to INTELSAT as an institution. Throughout, attention will be directed not only to the organization itself but also to the pattern of interplay between INTELSAT and its individual members, as well as among its members as individual nation-states.

Section A treats the issue of satellite systems separate from INTELSAT, their coordination procedure under Article XIV of the intergovernmental agreement and the considerations both for and against allowing such competition. Section B reviews the challenge of fiber-optic cables. Section C discusses the important role that the Communications Satellite Corporation has played in INTELSAT's evolution. It notes various difficulties encountered by Comsat in a domestic U.S. regulatory context and by INTELSAT as the result of Comsat's continuing influence in that organization. In Section D the deregulatory trends in the telecommunications sectors of the major industrialized members of INTELSAT are reviewed and then related to their effects on INTELSAT. Section E probes the effects on INTELSAT of the recent scarcity of launch vehicles. INTELSAT's status as an international organization and its relationships with other international organizations, particularly, the ITU, provide the subject matter for Section F. Finally, Section G analyzes INTELSAT's influence on the economically developing regions of the world, both from the standpoint of INTELSAT's specific programs and policies and from the broader perspective of telecommunications as a vital factor in the process of economic development.

A. *Separate Satellite Systems and Article XIV of the Intergovernmental Agreement*

No issue affecting INTELSAT has been more voluminously or passionately debated during the past five years than that of separate systems. Once anathema to the organization, separate systems are now being accepted by most INTELSAT members as inevitable technical and economic features of tomorrow's telecommunications landscape. There is every indication that at least some of the private U.S.-owned international systems will successfully withstand the scrutiny of INTELSAT's Board of Governors and Assembly of Parties under Article XIV(d) of the intergovernmental agreement.[1] This will mean that neither technical incompatibility nor »significant economic harm« to the INTELSAT system will be found to exist. This section seeks to place these recent and heavily reported developments in a historical and conceptual framework. In particular, Subsection 1 reviews the more straightforward requirements of Article XIV for domestic and specialized systems that involve only technical coordination. Subsection 2 presents the history of Article XIV(d) coordinations for systems such as EUTELSAT, PALAPA, ARABSAT and U.S. transborder networks. It also documents the regulatory history of the proposals by private U.S. firms. Subsection 3 summarizes the major arguments for and against the approval of recently proposed private systems. The future of separate systems is discussed in Subsection 4, while Subsection 5 contains some preliminary conclusions.

1. *Technical coordination of separate domestic and specialized systems under Article XIV(c) and XIV(e)*

Article XIV(c) of the intergovernmental agreement requires INTELSAT members establishing separate satellite systems for »domestic public telecommunications services« to »consult« with the Board of Governors beforehand. The latter is then to »express, in the form of recommendations, its findings regarding the technical compatibility of such facilities and their operation with the use of the radio frequency spectrum and orbital space by the existing or planned INTELSAT space segment.«

1 At its April 1987 meeting, the Assembly of Parties decided to allow five transponders of the PanAmSat system to begin leased, non-switched international service with Peru. Expansions of the system must be coordinated separately. This is the first successful Article XIV(d) coordination of a system that is both privately owned and truly international, as opposed to regional or transborder. Other coordinations under Article XIV(d) at this meeting include the following: (1) a teleport project in Jamaica funded by an international consortium of carriers, including American Satellite and NTT; (2) video service only for the French satellite TELECOM I to Mauritius, Madagascar and Haiti; (3) extension of Comsat's MARISAT system for use by INMARSAT; (4) extension of INTERSPUTNIK's existing traffic to Algeria and inauguration of INTERSPUTNIK traffic to Vietnam; (5) use of a U.S. satellite system for service to St. Pierre and Miquelon, two French islands offshore from eastern Canada; and (6) 21 U.S. systems for miscellaneous transborder traffic.

Table 1 – Technical Coordination of Separate Satellite Systems Under Article XIV(c) and XIV(e)

Year	Countries	Paragraph	Systems
1976	Brazil	XIV(c)	[Not indicated in document]
	France	"	OTS
	Japan	"	GMS
	United States	"	USA 2A, 3B, Marisat
	France	XIV(e)	Symphonie
	Italy	"	SIRIO
1977	Canada	XIV(c)	TELESAT B-1, C-1, C-2
1978	Colombia	XIV(c)	SATCOL 1, 2
	Indonesia	"	PALAPA-B
	Iran	"	ZOREH 1, 2, 3
	United States	"	USASAT 6A, 6B
1979	ARABSAT Countries	XIV(c)	ARABSAT I, II
	United States	"	WESTAR III
1980	Canada	XIV(c)	TELESAT D-1
	India	"	INSAT 1-A, 1-B
	Japan	"	CS-2a, 2b
	Malaysia	"	PALAPA A1, A1, B1, B2, B3
	Philippines	"	PALAPA A1, A2, B1, B2, B3
	Saudia Arabia	"	SABS
	Thailand	"	PALAPA A1, A2, B1, B2, B3
	Germany (Fed. Rep.)	XIV(e)	TV SAT
	India	"	INSAT 1-A, 1-B
	Japan	"	GMS-2
	Saudia Arabia	"	ARABSAT I, II; SABS
1981	Australia	XIV(c)	AUSSAT I, II, III
1982	Canada	XIV(c)	TELESAT C-3, D-2
	Japan	"	CS-2
	United Kingdom	"	UNISAT-1
	Australia	XIV(e)	AUSSAT I, II, III
	Italy	"	SARIT
	Japan	"	BS-2
	Luxembourg	"	LUXSAT
	Switzerland	"	HELVESAT-1
	United Kingdom	"	UNISAT-1
1983	Brazil	XIV(c)	SBTS A-1, A-2
	Colombia	"	SATCOL 1A, 1B, II
	France	"	TELECOM 1A, 1B, 1C
	United Kingdom	"	UNISAT
	United States	"	Advanced WESTAR I, II
	United States	"	COMSTAR D-4
	United States	"	GALAXY I, II
	United States	"	GSTAR I, II
	United States	"	SBS I, II, III
	United States	"	SPACENET I
	United States	"	TELESTAR 3A, 3B
	United States	"	US SATCOM I-R, II-R, III-R
	United States	"	US SATCOM IV
	United States	"	WESTAR IV, V
	Japan	XIV(e)	GME-3

TABLE 1 -- [continued]

Year	Countries	Paragraph	Systems
1984	Germany (Fed. Rep.)	XIV(c)	DFS 1, 2
	Israel	"	AMS 1, 2
	Italy	"	ITALSAT
	Mexico	"	MORELOS I, II
	United States	"	ASC-1
	United States	"	SBS 4 (USASAT 9A)
	United States	"	[18 networks for domestic use]
1985	India	XIV(c)	INSAT 1B, IC, D
	Italy	"	SARIT
	Norway, Sweden	"	TELE-X
	United States	"	ASC II
	United States	"	RCAA SATCOM K-1
	United States	"	SPACENET III
	United States	"	TELSTAR 302
	France	XIV(e)	TDF-1
	India	"	INSAT 1C
	Norway, Sweden		TELE-X

Sources: INTELSAT documents MS-6-10E, April 1977, pp. 17–18;
MS-7-10E, March 29, 1978, p. 20;
MS-8-10E, March 14, 1979, p. 21;
MS-9-10E, March 18, 1980, p. 23;
MS-11-10E, March 20, 1981, p. 22–23;
MS-12-10E, April 1, 1982, p. 22;
MS-13-10E, March 24, 1983, pp. 24–25;
MS-14-10E, (Rev. 1), April 4, 1984, p. 21;
MS-15-10E, April 2, 1985, pp. 23–24;
MS-16-10E, March 12, 1986, pp. 27–28.

Similarly, Article XIV(e) mandates that INTELSAT members establishing separate systems for »specialized telecommunications services..., domestic or international,« shall »furnish all relevant information to the Assembly of Parties, through the Board of Governors.« The Assembly of Parties, »taking into account the advice of the Board of Governors,« then makes recommendations similar to those for separate domestic systems.

Table 1 lists those INTELSAT members that, during the period 1976–85, coordinated separate systems under Article XIV(c) and XIV(e). A wide range of countries and systems is involved. Some systems, having properties of both domestic and specialized service, were coordinated under each provision. Others, such as PALAPA and ARABSAT, eventually entered into international public service and were later coordinated under Article XIV(d) as well. Many, at least in their earlier stages, were essentially experimental. These include the French-German Symphonie and the Italian SIRIO systems.

Others, such as LUXSAT of Luxembourg and the U.S. SBS system, were scaled down considerably from their originally projected capacity. This procedure of technical coordination for separate domestic and specialized telecommunications satellite systems, since it does not focus on economic features, has been relatively straightforward and uncontroversial.

2. *Economic and technical coordination of separate international systems under Article XIV(d)*

a. *Previous coordination procedures*

Beginning in 1979, several separate international systems were successfully coordinated with INTELSAT under the procedures of Article XIV(d). These are indicated in Table 2. Such procedures, it will be recalled, require a showing that »significant economic harm« will not be inflicted on INTELSAT, as well as assurances of technical compatibility.

By far the most significant of these systems, both in terms of their extent and capacity and their precedential significance for later systems, were the various configurations of the EUTELSAT system, sponsored by the EUTELSAT organization in Western Europe; the PALAPA system, owned by Indonesia and leased to neighboring countries; and the ARABSAT system, shared by various Arab states in the Middle East. A second identifiable group of such systems provides transborder services between adjacent or neighboring countries. These systems comprise a large number of networks serving traffic between the United States and Canada, Bermuda, the Caribbean, Mexico, and Central and South America that have been coordinated according to Article XIV(d).

All of these systems, however, as Caplan (1986, p. 196) notes, are regional rather than truly international in character. Proponents of each were able to make a convincing case that little or no traffic would be diverted from INTELSAT. In doing so they contended that in the absence of regional satellite systems the traffic in question would have been carried by terrestrial means rather than by INTELSAT. It is instructive in this regard to examine the findings of the Board of Governors in 1979 with respect to possible economic harm to INTELSAT from the establishment of the first EUTELSAT system:

> [I]n the interest of fostering European industrial effort in space development, and recognizing ... special circumstances which are unique to the European community, a number of European governments have agreed to provide a substantial subsidy to

Table 2 – Economic and Technical Coordination of Separate International Systems Under Article XIV(d)

Year	Countries	Systems
1979	France	ECS (EUTELSAT-I)
	Indonesia	PALAPA B1, B2, B3
1980	Algeria	Algeria/INTERSPUTNIK
	France *et al.*	EUTELSAT 1-2
	Indonesia *et al.*	PALAPA A1, A2
	Saudi Arabia *et al.*	ARABSAT I, II
1982	Canada and United States	[20 networks]
	France *et al.*	EUTELSAT I-1, I-2
	United Kingdom and United States	US SATCOM III-R and IV [TV in Bermuda]
	United Kingdom *et al.*	MARECS [for INMARSAT]
	United States *et al.*	MARISAT [for INMARSAT]
1984	France *et al.*	EUTELSAT I-3
	United States and Canada	ANIK B, C1, C1, C2, C3, D1, D2, ASC-1, SBS 4 (USASAT 9A)
	United States and Canada	[18 networks--international use]
	United States and Mexico	GALAXY(I (USASAT 11D)
	United States *et al.*	[18 networks--Latin American/Caribbean use]
1985	United States and Canada	ASC II; RCAA SATCOM K-1; SPACENET III; TELSTAR 302
	United States *et al.*	[23 networks--Mexico, Caribbean, South and Central America]
1986*		

Sources: INTELSAT documents MS-9-10E, March 18, 1980, p. 24;
MS-11-10E, March 20, 1981, p. 23;
MS-13-10E, March 24, 1983, pp. 24–25;
MS-15-10E, April 2, 1985, pp. 23–24;
MS-16-10E, March 12, 1986, p. 28.

* For information on Article XIV(d) coordinations through April 1987, see footnote 1 of this chapter.

> support the development and implementation of the ESC [EUTELSAT] system to assure its use as a supplement to the terrestrial network. It is by virtue of this subsidy that the traffic which, on a normal telecommunications approach would be carried terrestrially, can be carried by the ESC system. The project is not intended to impact significantly on the general character of the European network as a terrestrial one .... With few if any exceptions, the circuits to be carried by the ESC system would, in its absence, have been carried by the European terrestrial network and not by INTELSAT (BG-37-54E, March 1979, p. 6).

Similar financial and operational considerations were at work in the ARABSAT system as well. The PALAPA system, however, was originally designed as a domestic system for Indonesia; later, a decision was made to lease spare capacity to the neighboring countries of Malaysia, the Philippines, Singapore and Thailand. Representatives of PALAPA convinced the Board of Governors that since the system »will be confined to traffic originating in remote locations in the participating countries« and »the small volumes of traffic from the remote locations would make ... alternative solutions ... uneconomic or impractical,« »it is unlikely that any significant proportion of the ... traffic that the system will carry would flow on the INTELSAT system in the foreseeable future« (BG-37-53E, March 1979, p. 5).

b. *Recent proposals for separate international systems by private U.S. firms: a chronology*[2]

Beginning in 1983, a number of U.S. satellite entrepreneurs formed ventures aimed at establishing international satellite systems separate from INTELSAT. Accordingly, they approached the Federal Communications Commission for preliminary approval. The original six firms are: Orion Satellite Corporation (Orion), proposing non-tariffed, leased or sold transponder capacity between Western Europe and the eastern United States and Canada; International Satellite, Inc. (ISI), proposing both tariffed common-carrier services and leased and sold transponders between the continental United States and Western Europe; Cygnus Satellite Corporation (Cygnus), proposing leased or sold transponder capacity between the continental United States and Western Europe, with a spot beam to the Caribbean and portions of Central America; Pan American Satellite Corporation (PanAmSat), proposing the sale or long-term lease of transponders for international traffic between North and South America and for domestic service within South American countries; RCA American Communications, Inc. (RCA), proposing service between the continental United States and portions of Europe and Africa on a tariffed, common-carrier basis as well as by means of whole transponder leases; and Financial Satellite Corporation (FINANSAT), proposing interconnected satellites over both the Atlantic and Pacific Oceans offering transponders for sale or long-term lease to major business on a non-common-carrier basis. Later, two Pacific systems entered the race. These were Columbia Communications Corporation (Columbia), proposing to link the western continental United States, Alaska, Hawaii, western Canada and Japan with transponders for sale or long-term lease; and McCaw Space

2 All events in Subsection 2(b), including those not specifically footnoted, are documented in U.S. Federal Communications Commission (1986, pp. 5–7).

Technologies, Inc. (McCaw), proposing satellites over the Indian and Pacific Oceans interconnected with a telecommunications linkup in Guam and offering services throughout the United States, the Pacific rim and basin, Asia, the Middle East and parts of Europe and Africa on a non-common-carrier basis.

As is often the case when the FCC faces important new policy challenges, the U.S. president issued a directive intended to signal the preference of the executive branch of the federal government.[3] On November 28, 1984, President Reagan signed a presidental determination (U.S. President 1984) stating that, according to provisions of the Communications Satellite Act of 1962, »separate international communications satellite systems are required in the national interest.« The United States government was to »consult with INTELSAT regarding such separate systems as are authorized by the Federal Communications Commission.« On the same day, the Secretary of State, George P. Shultz, and the Secretary of Commerce, Malcolm Baldridge, issued a joint letter to Mark S. Fowler, the FCC Chairman, requiring that the following conditions be satisfied prior to final FCC authorization »to assure that the United States meets its obligations as a party to ... INTELSAT«:

> (1) each system is to be restricted to providing services through the sale or long-term lease of transponders or space segment capacity for communiations not interconnected with public-switched message networks (except for emergency restoration service); and,
>
> (2) one or more foreign authorities are to authorize use of each system and enter into consultation procedures with the United States Party under Article XIV(d) of the INTELSAT Agreement to ensure technical compatibility and to avoid significant economic harm (U.S. Department of Commerce 1984).

On July 25, 1985, the FCC issued its *Separate Systems* decision (U.S. Federal Communications Commission 1985), establishing policies and procedures to consider applications for separate systems. On the same day, it granted provisional construction and operation authority to ISI, PanAmSat and RCA. Later it granted similar permission to Cygnus, FINANSAT, Columbia and McCaw. In doing so, it rejected requests by the Communications Satellite Corporation and other firms to restrict or overturn these authorizations. In reaching this decision, the Commission reiterated its determination not to grant launch authority or permit actual system operation »until the United States has completed consultation with INTELSAT pursuant to Article

3 For an account of the White House initiative establishing competition in the U.S. domestic satellite industry, see Oslund (1977).

XIV(d) ... and we have been informed by the Department of State that the United States has fulfilled its obligations under Article XIV(d)« (U.S. Federal Communications Commission 1986, p. 7).

Successful coordination with INTELSAT, in turn, requires that a system name at least one sending authority in one country and at least one receiving authority in a different country. Following the issuance of preliminary FCC approval, this became the next regulatory hurdle for the field of private firms competing to establish separate systems. Orion was the first firm to approach the FCC and to attract substantial attention from the mass media and the trade press. Peru's recent agreement to participate with PanAmSat, however, makes PanAmSat the only proposed system to date with a firmly committed foreign correspondent. In March and early April 1987, PanAmSat, along with several other systems, was successfully coordinated under the Article XIV(d) procedure with the Board of Governors and Assembly of Parties (*see* footnote 1). Despite its earlier monolithic opposition to separate systems, INTELSAT now seems inclined to yield to the inevitable. Given the favorable outcome, PanAmSat is reportedly preparing to launch its first satellite during the summer of 1987 using an Ariane rocket.

3. *Arguments for and against the approval of separate systems*[4]

As might be imagined, proponents of both the status quo and of entry by private separate systems have been effective and prolific advocates of their individual positions. Under the energetic leadership of its director general from 1983 to 1986, Mr. Richard R. Colino, INTELSAT mounted an unprecedented campaign to oppose separate entry. In doing so, it hired lobbying firms and engaged members of Congress, the executive branch and the FCC in dialogue (*see*, *e.g.*, Colino 1983). Apart from the United States, INTELSAT at first opposed the entry of separate systems with virtual unanimity. Arrayed against these efforts were the studies and policy statements of the executive branch, primary among which was the White Paper on New International Satellite Systems (U.S. Senior Interagency Group 1985), issued early in 1985 by a Senior Interagency Group composed of representatives of the executive branch. The satellite firms themselves likewise launched well-funded lobbying efforts and engaged in regulatory proceedings

4 Arguments for and against separate systems are developed at length in Snow (1985a), on which Subsection 3 draws.

before the FCC that, as we noted above, were by and large successful (*see, e.g.*, Pan American Satellite 1986).

It is not the purpose of this study to reproduce these arguments at length or in depth. Indeed, for reasons we shall set forth later, the approach of separate systems has acquired an almost inexorable technological and economic dynamic that appears to have pierced the formidable defenses that INTELSAT and most of its members had erected against it. The question now is not whether separate systems will come but rather when and to what extent they must be reckoned with and how INTELSAT can best position itself to remain a healthy and viable force in international telecommunications after several such systems are in operation.

In retrospect, the argumentation both for and against separate systems has missed the point. Article XIV(d) specifically requires a finding of whether »significant economic harm« will result from the entry of separate international systems. Despite this requirement, much of the reasoning on either side has been political in nature. Such economic arguments as have been put forth have lacked both methodological rigor and foundation in economic theory.[5] One observer, having surveyed this field, concludes that »[t]he choice is ... wholly political« (Potamitis 1986, p. 54). Another states that »in political terms light still takes years to circle the earth. The problem is not simply one of technology, nor is it one of economics. The problem is one of world politics« (Caplan 1986, p. 201).

The purpose of the rest of this subsection is to supplement, if only retrospectively, these essentially political and adversarial arguments with a distillation of the most important economic, institutional and philosophical considerations for and against separate systems that could have been put forward by each side – and to some extent actually were.

a. *Arguments against separate systems*

By far the most potent argument against allowing the entry of separate systems is the possibility that INTELSAT, assuming it has sufficient economies of scale and scope, is a natural monopoly. As we have seen, this property would imply that INTELSAT could produce certain (or perhaps all) combinations of output more cheaply than could two or more competing satellite

5 Snow's (forthcoming 1987b) study is the only econometric treatment of the issue to date. As noted, it concludes that INTELSAT's cost function exhibited natural-monopoly properties during the period 1967–1984.

enterprises. In some sense, there would be a loss to society if the least-cost output configuration did not emerge. This can happen, it will be recalled, if the natural monopoly is unsustainable; in that case, there are financial incentives for smaller, more specialized firms to enter and to succeed in capturing some of the incumbent monopolist's market. It is easy, however, to beg a number of important questions by being vague about the »markets« served and the »services« provided by the incumbent firm and its latent competitors. The private firms have uniformly contended that they will serve traffic that INTELSAT is not equipped to handle: non-switched transponder leases and sales outside of the public network. We shall return to this point later.

INTELSAT has declared itself amenable to allowing service competition in place of facilities competition. It has moved in that direction in recent years by offering an increasing number of long-term transponder leases in its tariffs without restriction as to the resale of this capacity to third parties or its enhancement and repackaging in the process. Jonscher (1986) has related the dichotomy between facilities and service competition to a similar one between horizontal and vertical competition in the British context.

INTELSAT's average-cost and non-discriminatory pricing policy is an additional argument used against the entry of separate international systems (*see* Snow 1979). Such systems are seen as capable of »cream skimming« by concentrating on lucrative, high-volume, low-cost routes while INTELSAT incurs losses on its thinner links. Indeed, it has been argued that INTELSAT's prices, when they are set at average cost, send perverse signals to entrants (*see* Johnson 1987 *passim*). Since INTELSAT's average cost is above its marginal cost, it is argued, an entrant can succeed in siphoning revenue from INTELSAT by charging at its own marginal cost even if that is above INTELSAT's marginal cost (but still below INTELSAT's average cost). The simple answer to these objections is that INTELSAT's pricing statutes and behavior must be changed to allow it to compete and survive in a post-entry world. Our review of the INTELSAT tariff in Section A of Chapter III presents ample evidence that INTELSAT is already interpreting its pricing mandate with considerable flexibility and imagination. Nevertheless, Article V(d) of the intergovernmental agreement and Article 8 of the Operating Agreement, if suitably amended to allow greater latitude for welfare-optimizing pricing practices, would help to eliminate incentives for cream skimming and »uneconomic« entry by competitors.

Habit, preference and risk aversion by consumers and other users are often cited as reasons for opposing change in a telecommunications environment. Lamberton (1986), for example, has made this point with respect to Australia.

Likewise, Noam (1986b) has discussed such phenomena at length in the context of European postal and telecommunications administrations. It may be, for example, that a substantial measure of price stability and predictability is preferred by most users to prices that over time exhibit a declining mean with a high variance. Much of INTELSAT's monolithic opposition to separate systems may derive from the fact that the majority of its signatories are telecommunications administrations with various organizational and psychological motives for opposing change. These motives may very well have validity in terms of human and material resources, and they should be evaluated and included to the extent possible in the overall calculus of costs and benefits leading to a choice among alternative policies.

A very common argument against the entry of separate systems is that it would force INTELSAT to price in a manner that would eliminate the considerable subsidy to developing areas now implicit in its rate structure. This argument has taken on added weight since the entry into force of the definitive agreements, which have accorded greater influence to the views of small users. From an economic point of view, the best reply is that subsidies implicit in a pricing system are a relatively inefficient way of redistributing income and wealth. Direct grants, or other explicit subsidies, are much more effective in attaining the same goal, although their very transparency makes them politically sensitive and difficult to agree upon. As we shall observe in Section G of this chapter, INTELSAT already sponsors a number of programs that directly benefit Third World countries. Increasing this type of assistance by direct external funding would appear to be a more effective way of aiding the process of economic development than perpetuating an economically inefficient implicit subsidy by means of overly rigid (or too rigidly interpreted) mandates in the definitive agreements.

The cooperative nature of INTELSAT's organization and financial structure is one of its enduring and appealing features. Many have argued that cooperation, rather than competition, should be the hallmark of the telecommunications sector. The French telecommunications authorities (*see* Voge 1986; Bruce, Cunard, and Director 1986, pp. 505–40), for example, have stressed cooperation (and enthusiastic entrepreneurship practiced by PTT subsidiaries) as superior to competition.

Social economy *(Gemeinwirtschaftslehre)* is a German variant of this line of thought (*see* Snow 1984; Thiemeyer 1983). It derives from nineteenth-century German economic thinkers and the English cooperative movement. It likewise emphasizes the need for a single, public provider of the most important sectors of the economy: communications, power, water, transportation, education, health and so on. While arguments favoring cooperation over competition are seldom couched in an analytical or mathematical framework, they often have a persuasive rhetorical and ideological ring that is quite influential in buttressing arguments against deregulation, privatization and competitive entry in the telecommunications sector, particularly in Western European countries.

Finally, foreign-policy considerations have played a role in the separate-systems debate, albeit a clearly minor one. Initially, INTELSAT's argument went along these lines: why should the United States, which was instrumental in establishing INTELSAT, in providing its technology and personnel and in guiding it through its early years, now desert the organization just as it reaches maturity? Why should it oppose the wishes of 112 of its 113 members? The U.S. Department of State, however, closed ranks with the rest of the executive branch and did not show itself particularly sympathetic to this line of thought (*see* U.S. Senior Interagency Group 1985, pp. 18–29). Instead, it subscribed to the strongly procompetitive and deregulatory thinking that characterized the first Reagan administration. It should not be supposed that this decision was based on unthinking conservatism or ideological zeal. Instead, quite careful analysis within the executive branch led the administration to the conclusion that the costs of supporting separate entry – and foreign relations figured prominently among these – were exceeded by the economic benefits of competition. These will be outlined now.

b. *Arguments in favor of separate systems*

Perhaps the most cogent argument in favor of separate systems can be couched in a dynamic framework. As time passes, new technologies respecify the cost functions on which the static properties and assumptions of natural monopoly are based. For reasons that are still unclear, a competitive environment, at least in the telecommunications sector, most often seems to be the best one in which service providers can respond quickly and flexibly to the broad and growing range of user needs. This was the basis of the U.S. decision in 1970 to allow competition in domestic satellite systems (*see*, *e.g.*, Snow 1977a); it seems unchanged in its essence as a justification of the more recent FCC decision to support separate international systems. As has been noted (*see* von Weizsäcker 1986), a dynamic framework is compatible with the kinds of cost-based pricing that can eliminate cream skimming and other economically inefficient cross-subsidies, while it is the static framework that is more amenable to demand-based pricing schemes such as price discrimination and Ramsey pricing. These latter methods, as we have seen, are not only proscribed by INTELSAT's agreements but also present serious methodological and political obstacles to their implementation as well.

A second argument in support of separate systems has to do with the markets being served by INTELSAT and its potential competitors. The United States government asserts, with considerable justification, that it has protected INTELSAT from »significant economic harm« by requiring entrants to provide their services through »the sale or long-term lease« of capacity »not interconnected with the public-switched networks.« Only ten to twenty percent of INTELSAT's present traffic, however, consists of such services; the rest is either switched public traffic, short-term leased traffic or both. Each potential entrant has argued that it is not allowed, as a result of these requirements, to compete for the »core« of INTELSAT's traffic. There is, to be sure, legitimate concern about the extent to which regulations alone can prevent users from »leaking« part of their traffic into the public switched network without fear of detection, given today's technology.[6]

Reference to a »core« of INTELSAT's service offerings leads to a third consideration favoring entry of separate systems. The interagency White Paper refers to »an essential core for public-switched international communications« (U.S. Senior Interagency Group 1985, p. 26), and the FCC expresses a belief that its entry policy is sufficiently restrictive to avoid »the threat of significant economic harm to INTELSAT's commercial core« (U.S. Federal Communications Commission 1986, p. 33). Economically, this core is the natural-monopoly portion of INTELSAT's service offerings – that part of its output that INTELSAT's single global system can offer more cheaply than can any combination of two or more systems. How might this natural monopoly core be discovered? There would appear to be some consensus that it is more or less coextensive with INTELSAT's public switched voice offerings, but no empirical work has been undertaken to determine whether that is the case. Indeed, problems such as this in the definition of relevant markets are a common and difficult topic in the legal and economic literature relating to competition policy in industrialized countries.

Following Hayek's (1968) views on competition as a decentralized, autonomous and efficient »discovery procedure« or information system, it can be argued that competition by separate systems with INTELSAT is the most efficient method of »discovering« whether INTELSAT has a natural mono-

6 »A PBX in Chase Manhattan's New York offices could, for example, take an incoming call from one of the bank's European branches over a leased Orion circuit and switch it onto the local phone system in order to terminate the call at a customer's factory in Albany« (Cowhey and Aronson 1985, p. 31).

poly core, and, if so, what its extent is. This core will simply be that portion of INTELSAT's traffic that rival entrants are unable to compete away (we assume the absence of predatory pricing, subsidies and other distortions of legitimate price signals, enforced by appropriate antitrust provisions if necessary). Such a »discovery« will, it is argued, be made more rapidly, accurately and at a lower cost under competition than it would be by either direct economic analysis or bureaucratic fiat.

Two examples suggest themselves. As consultants for West Germany's Monopolies Commission, Knieps, Müller and von Weizsäcker (1981) argued that the Deutsche Bundespost should not be precluded a priori from competing with private firms in supplying the terminal-equipment market. Such competition, they maintained, would »discover« the extent (if any) to which the Bundespost enjoyed economies of scope in its joint provision of network and terminal-equipment services. Likewise, the definition and establishment of local access and transport areas (LATAs) in the United States following the divestiture of the Bell system appears to be a similar attempt – although using somewhat arbitrary administrative methods rather than market forces – to define spatially disparate core areas of natural monopoly in the public switched voice market (*see, e.g.*, Noll 1986a).

There is one rather theoretical objection to allowing market forces free rein in determining the extent of natural monopoly. As we have seen, an unsustainable natural monopoly furnishes financial incentives for what might be called »uneconomic« entry. If an entrant is successful in competing away part of an incumbent's business, this might be interpreted either as evidence of a lack of natural monopoly on the part of the incumbent or as evidence that the incumbent exhibits an unsustainable natural monopoly. A way out of this dilemma is suggested by the fact that sustainability is largely a function of the pricing strategy used by the incumbent firm. We have noted, for example, that a strictly Ramsey-priced tariff, despite the political and practical obstacles to its implementation, can assure sustainability under a broad range of conditions. There is a good deal of additional conceptual and empirical research to be done in this area. Nonetheless, there appears to be considerable merit in the notion of competition as a way of gaining access quickly and efficiently to information regarding underlying cost conditions. This was clearly a motivation for introducing competition into domestic U.S. satellite markets, for example.

A fourth argument in favor of separate systems concerns the financial, administrative and institutional dichotomy between INTELSAT's space segment and its ground segment. The latter, as we have seen, is not owned or controlled by INTELSAT but rather by its 113 signatories. A best guess as to the share of INTELSAT space-segment costs in total user-to-user costs is that it does not exceed ten or fifteen percent. Representatives of separate U.S. systems attempting to coordinate with INTELSAT have asked why more stress is not placed on reducing costs in the portion of the satellite link that accounts for the overwhelming portion of total costs. The telecommunications ministries in INTELSAT's Meeting of Signatories often have financial and institutional incentives to keep these costs high by configuring the terrestrial infrastructure more densely than is economically optimal, as Caplan pointedly suggests:

> The power brokers in European satellite communications know full well that [separate satellite systems will not destroy INTELSAT]. However, they also know that separate systems will, as part of their service, transmit to small earth stations owned by and located at the end user's place of business. This service will result in the bypassing of the terrestrial network which, in the past, has been used to carry signals from the large INTELSAT earth station to the end user. And thus, the PTTs who own and operate the terrestrial networks will lose on intracorporate communications. Until now, the PTTs have had little to say about this possibility, preferring to hide behind the shield of INTELSAT (Caplan 1986, p. 197).

A fifth argument in favor of competition is quite straightforward: even a credible threat of entry, not necessarily its reality, constrains the incumbent to behave in a fashion more in the interest of consumers and to set prices closer to costs. Jonscher (1986) has noted this phenomenon with respect to the behavior of British Telecom (BT) since its exposure to competition by the Mercury consortium. Others have suggested that much of INTELSAT's recent proliferation of new service offerings, concessionary tariffs and relaxation of rigid earth-station standards coincided with, and may have been introduced to forestall, the threat of separate systems that began with the Orion initiative in 1983. The recently proposed theory of contestable markets, as outlined in Section C(6)(d) of Chapter I, provides a strong conceptual foundation for the heuristic appeal of this line of thought.

A sixth reason for allowing competition, one that is advanced by large international business users in particular, concerns the institutional ease with which private systems can be designed and operated. One of the appeals of submarine cable systems, as we shall see, is that they emerge from quiet, often bilateral negotiations among the administrations involved. Rightly or wrongly, large business users are coming to view INTELSAT as a cumbersome and increasingly politicized international organization that is insensitive to the needs of corporate customers. They view the »administrative bypass« of such an organization, and the ability to deal quietly and effectively with only a small number of administrations, as desirable aspects of private separate systems.

Optimal pricing to promote overall economic welfare[7] is a seventh consideration cited in favor of separate systems. By forcing prices closer to costs, competition emits economic signals and incentives for a more nearly optimal distribution of productive resources. We have noted, for example, that incentives for »cream skimming« and other inefficient patterns of subsidy disappear as prices are forced closer to costs. The pricing adjustments necessitated by introducing competition are by no means free of political implications, however, as the deregulatory experience in the United States (*see* Baughcum 1986) and elsewhere has indicated.

Finally, proponents of separate systems have claimed that in the present context of exponentially increasing satellite usage, competition for certain major markets might actually increase INTELSAT's own traffic, other things equal (Fishman 1984, p. 112). In its less extreme version, this argument runs to the effect that the historically high rate of absolute traffic growth quickly compensates INTELSAT for any relative losses suffered as a result of competitive encroachments.

The basis in theory of this alleged phenomenon is not well articulated. The contention implies that the services offered by competing satellite carriers are somehow complements rather than substitutes for INTELSAT's services. Users leasing transponders from a small private carrier for specialized video or data services, for example, might find their public switched telephony requirements for the same route (i.e., those provided by INTELSAT) increased as a result.

The validity of this line of reasoning depends to a considerable extent on assumptions about the market niche being targeted by competitors. This might include services that INTELSAT (a) has no intention or ability to offer; (b) would offer only if they were also provided by competitors; (c) would offer only if there were a credible threat of their being provided by competitors; or (d) would offer in any event, regardless of the intentions or actions of possible competitors. In the absence of information about the markets involved, and of theoretical models setting forth behavioral assumptions and hypotheses that can be empircally tested,[8] the claim that competitive offerings on INTELSAT's routes would stimulate rather than divert or depress INTELSAT's traffic is still quite speculative.

7 For an annotated bibliography of pricing theory and tariff policy, see Snow and Jussawalla (1986, Chapter 5, pp. 58–66).

8 Sharkey's (1982, pp. 111–44) game-theoretic approach to destructive competition is instructive in this regard.

4. *The future of separate systems*

As the foregoing subsection has noted, there is an increasing consensus that the emergence of at least a modest number of separate systems is by now a foregone conclusion. This is the result of relentless technological and economic forces in telecommunications technology. In retrospect, perhaps too much attention has been devoted to the identities and behavior of the eight or so private U.S. systems that are now searching for foreign correspondents and awaiting Article XIV(d) coordination after having received preliminary FCC approval. What will happen in the likely event that at least some of these systems are actually launched and begin operation?

First, it is probable that few of these systems can remain financially viable if they are limited to long-term transponder sales and leases outside the public switched network. This limitation was imposed by the FCC to protect INTELSAT from economic harm; it was not part of the original business plan of the private entrepreneurs. Once these firms have found non-U.S. correspondents (and only PanAmSat has to date), undergone Article XIV(d) coordination and begun operation, many of them will seek to enlarge the scope of their operations to include the public switched network, short-term leases and as many additional non-U.S. countries as possible.

As such systems begin to proliferate, INTELSAT may begin to consider the »cumulative« economic harm accruing from all systems together rather than the »marginal« economic harm caused by each one separately.[9] This raises two important questions regarding the force and effect of Article XIV(d): does it *forbid* the establishment of a separate international system found to cause »significant economic harm«; and regardless of whether it does, would the United States government support such a finding by refusing to grant launch and operation authority to a system unsuccessful in obtaining Article XIV(d) coordination from INTELSAT?

For its part, the United States government has consistently asserted that Article XIV(d) does not commit INTELSAT members to refrain from establishing separate international systems in the event of an affirmative finding with respect to the presence of significant economic harm (*see* U.S. Senate 1983, p. 144). To underscore this, it has spoken of »consultation« regarding the provisions of Article XIV, while INTELSAT and most of its

9 The issue of cumulative as opposed to marginal economic harm was addressed by the Assembly of Parties in 1985. The Assembly stated that it would »take into consideration the cumulative economic effect on INTELSAT of one or more systems submitted by a Party or Parties ... over an appropriate period of time« (MS-16-10E, March 12, 1986, p. 25).

members prefer to use the stronger term »coordination.« The language of Article XIV(d) is vague on this point, perhaps intentionally so. Even without binding power in the INTELSAT agreements to forestall the establishment of separate systems, however, it seems unlikely that the United States or other governments would risk flouting the wishes of more than 100 countries in the event that the Board of Governors and the Assembly of Parties found a high probability of significant economic harm. The following statement in the interagency White Paper addresses this issue:

> The United States is committed to ensuring that non-INTELSAT satellite systems are technically compatible with existing and planned INTELSAT satellites, and to avoiding significant economic harm to the global INTELSAT system. Accordingly, the Executive branch will initiate consultations with INTELSAT only for those non-INTELSAT systems which it believes meet the technical and economic conditions described in the INTELSAT Agreement. The United States will continue to consult with INTELSAT in good faith; therefore, the possibility cannot be excluded that, following the consultation process, the Executive branch might find that final regulatory authorization should not be granted.
>
> The consultation with INTELSAT would normally end with a recommendation being made by the INTELSAT Assembly of Parties. Such recommendations are not binding on the United States, although the U.S. Government will carefully consider all recommendations. It will go forward only with systems it deems consistent with its obligations to INTELSAT (U.S. Senior Interagency Group 1985, pp. 17–18).

In the event of U.S. entry into competition with satellites, it is to be expected that other industrialized countries would soon follow suit. Three countries – France, Japan and the United Kingdom – appear particularly well prepared for such initiatives. Many knowledgeable observers believe that the French TELECOM I satellite, launched in 1984, could be easily extended beyond its nominal designation as a domestic satellite. Spot beams aimed at overseas French departments offshore from eastern Canada, in the Caribbean and in the Indian Ocean, for exemple, could be reconfigured to serve French-speaking users in Quebec, Haiti and Madagascar, respectively.[10] Japan, for its part, proposed early in 1983 a two-tiered satellite system for urban and rural usage throughout the Pacific rim and basin.[11] British Telecom, as a newly

10 This point is made by Caplan (1986, p. 185); by Cowhey and Aronson (1985, p. 35); and in the Interagency White Paper (U.S. Senior Interagency Group 1985, p. 25).

11 *See* Research Institute for Telecommunications and Economics (1983) and Cowhey and Aronson (1985, p. 35). RITE's outline of the system contained considerable technical detail but was conspicuously vague about financial and institutional arrangements.

entrepreneurial telecommunications authority engaged in a global battle for markets with Cable & Wireless, would be another logical international entrant (*see* Cowhey and Aronson 1985, p. 35; U.S. Senior Interagency Group 1985, p. 25). Other possibilities include systems sponsored by Luxembourg, Ireland and Sweden (*see* Cowhey and Aronson 1985, p. 35).

5. *Conclusions regarding separate systems*

It appears likely, based on the information presented above, that the United States and perhaps certain other aerospace powers will soon establish international satellite systems separate from INTELSAT. The arguments both for and against entry, whether they are based on politics, technology or economics, seem moot and somewhat dated by now. The entry of separate systems is as much a symptom as it is a cause of the various economic threats to INTELSAT. The ongoing technological dynamic in telecommunications has not only overcome the scarcity and high cost of satellite channels but has also facilitated the provision of a much greater diversity of services to users (*see* Snow 1985b). As a result, the advent of separate systems has been rendered inevitable because of the restructuring of economic incentives and signals that the new technology has entailed.

There is an obvious but compelling analogy to be drawn here. INTELSAT is in many ways recapitulating the experience of the major carrier in at least three industrialized countries – the United States, the United Kingdom and Japan – that have consciously decided to liberalize their domestic telecommunications sectors. A competitive fringe has been allowed to carry specialized traffic that the monopoly firm has not served particularly well, while an effort has been made to preserve the core natural-monopoly traffic streams on which entry is presumed to be uneconomic. Each of these countries, as we shall see in Section D of this chapter, is undergoing considerable stress and strain in the process, but the overall outcome is a positive one. Neither AT&T, British Telecom nor Nippon Telegraph and Telephone (NTT) appears destined to dwindle into insignificance. Instead, each has been reanimated and strengthened by the threat and reality of competition. Given the will and resources to respond flexibly to its new challenges, there is little doubt that INTELSAT can similarly prevail against the challenge of competing systems.

B. *The Threat of Fiber-Optic Cables*

Despite the greater attention and regulatory effort devoted to separate satellite systems, well-informed individuals are virtually unanimous in their view that fiber-optic submarine cables pose a greater threat to INTELSAT's long-term financial and institutional viability. An entirely different technology is involved, and INTELSAT's agreements do not oblige its signatories to coordinate submarine cable systems, or any other systems using non-satellite transmission modes, to ensure that technical or economic harm to INTELSAT is avoided.

The satellite and cable modes are partial but not complete substitutes. Satellites have advantages due to their innate broadcasting ability and ease of interconnection, as well as their superior performance in terms of continuity of service. Many users, however, prefer cables because of their lack of propagation delay, their greater security from eavesdropping and certain technical features that make them better adapted to voice and data transmission. In addition, many national governments regard diversity of modal routing as a desirable feature for national security.[12] Thus, it is unlikely that one mode will completely replace the other. High-frequency radio systems, after all, are still in use in many parts of the world.

U.S. landing rights for at least five fiber-optic cables have already been requested from the FCC. Those with the earliest operational dates are the TAT-8 in the Atlantic and the HAW-4/TPC-3 in the Pacific, each to enter service in 1988 (Johnson 1987, p. 2). These cables will be based on the usual bilateral or multilateral negotiations among the telecommunications administrations involved, including AT&T on behalf of the United States. In addition, there are at least three private fiber-optic cable ventures in the offing. Tel-Optik, an undertaking financed by Cable & Wireless and by NYNEX, the post-divestiture Bell holding company in New York and New England, plans to lay cables in the Atlantic in 1989 and 1992 to transmit voice and data, respectively, between North America and Western Europe (Potamitis 1986, pp. 38–39). A venture called Submarine Lightwave Cable proposes to lay fiber-optic cables across the Atlantic (Bruce, Cunard and Director 1986, p. 273; Potamitis 1986, p. 39, note 48). Also, Pacific Telecom has proposed a fiber-optic cable from the state of Washington following the great circle route

12 The viewpoint of the U.S. Department of Defense is set forth by the U.S. Senior Interagency Group (1985, pp. 43–45).

to Japan, with a spur to Alaska (*see* Elliott 1987; Johnson 1987, p. 36). A Japanese partner for the venture has not yet been secured, however.

Cost comparisons between satellites and fiber-optic cables are extremely difficult to conduct because of the many definitional and other problems that are entailed. INTELSAT, for its part, has financed a study showing that under conservative assumptions (e.g., a ten-year satellite lifetime), the cost of a telephone circuit on a fully loaded INTELSAT VI spacecraft is $504, while the corresponding figure for the TAT-8 cable is $1,596.[13] There are at least two structural obstacles, however, that make it difficult to transfer these alleged cost savings – or any underlying cost advantages, whether they favor satellites or cables – to the user. First, the FCC's proportional-fill policy, requiring that AT&T use an approximately equal number of satellite and cable circuits to complete its overseas links, obviously prevents that firm from responding flexibly to relative cost conditions. This proportional-fill policy has its rationale in the belief that AT&T, as a rate-base regulated company, has an uneconomic incentive to use its own cables (which it can include in its rate base) rather than Comsat's satellite circuits (which it must lease and charge to expenses). Johnson,[14] however, argues that whatever bias toward cables AT&T might exhibit is not »strong enough to justify continuation of circuit-loading restrictions« (Johnson 1987, p. 39). Recent FCC policy has favored a slightly more flexible cable-satellite loading policy for AT&T, and the Commission will investigate the subject in depth by the end of 1988 (Johnson 1987, p. 3).

INTELSAT's average-cost charging policy poses a second and more fundamental obstacle to passing along any satellite cost savings through the price system. For example, INTELSAT is unable to exploit its economies of scope and scale on the dense Atlantic and Pacific streams when it prices on a route-by-route basis. Johnson has stated this dilemma succinctly for the case in which AT&T charges marginal cost and INTELSAT charges average cost:

> As long as excess capacity exists within each participant's share, each will (correctly) perceive the marginal cost of using an additional circuit as near zero. Thus, the price of adding a circuit is equal to marginal cost for cable, while the price exceeds marginal cost for satellite...

13 *See* Hampton (1986, p. 5). Bruce, Cunard and Director (1986, p. 288), however, assert that fiber-optic cables are »less costly.« Johnson's (1987) working hypothesis appears to be that the costs of both modes are roughly comparable.

14 It is noteworthy that Johnson (*see* Averch and Johnson 1962) is a co-author of the seminal paper that first established the theoretical explanation of overcapitalization by rate-base regulated firms. This behavior has come to be known as the Averch-Johnson effect.

> Given the way that costs are shared, carriers will have powerful incentives to substitute cable circuits for satellite circuits, although the marginal cost to the world for either may be about the same (Johnson 1987, p. 34).

We may summarize our conclusions regarding the threat of fiber-optic cables to INTELSAT as follows. First, technical differences in these two transmission modes, and the preference of many countries for a modal mix on national-security grounds, make it likely that both cables and satellites will survive into the foreseeable future. Second, there is no objective evidence to the effect that the fiber-optic cables to be laid in the late 1980s and early 1990s have innate cost advantages over the INTELSAT VI generation of satellites to be launched during the same period. Third, the FCC's proportional-fill policy and in particular INTELSAT's average-cost pricing policy may well prevent the pricing system that confronts the final user from accurately reflecting any underlying cost advantages of either mode. Finally, at least if the issue of pricing flexibility is not resolved, fiber-optic cables will pose a greater long-term threat to INTELSAT than will separate international satellite systems.

C. *The Role of Comsat in INTELSAT*

The present status and future prospects of INTELSAT are closely linked to the regulatory history and present business strategy of the Communications Satellite Corporation.[15] The first subsection below will present and comment on Comsat's recent regulatory history. The second will assess Comsat's policy options for the future and discuss the effects of such options on INTELSAT.

1. *Comsat's recent regulatory history*

In its *Authorized User II* decision, the FCC allowed Comsat to become a retailer by offering its services directly to the public (Sarreals 1986, p. 280). Until that time, it had been constrained by the *Authorized User I* decision to be a »carrier's carrier,« or wholesaler, by selling its circuits to AT&T and the

15 Historically minded readers might detect a certain resemblance to the role of Prussia in the establishment of the German Empire.

domestic record carriers that handled telex, telegraph and data traffic. In its 1985 *Authorized User III* decision, the FCC upheld its earlier ruling and spelled out provisions for its implementation. Among other activities, Comsat was authorized to offer INTELSAT capacity to all users at any U.S. earth station and to provide switched, leased-channel and other end-to-end services (Sarreals 1986, pp. 286–87). In its 1984 *Earth Station* decision, the Commission authorized individual international carriers to own U.S. earth-stations individually or jointly, thereby overturning its 1966 decision requiring Comsat and those carriers to share earth-station ownership in an Earth Station Owners Consortium (ESOC) in which Comsat held a fifty-percent ownership share and bore full management responsibilities as well. This same decision allowed Comsat to own and operate U.S. earth stations on its own through a separate common-carrier subsidiary (Sarreals 1986, p. 281). The *Earth Station* decision also required Comsat to disaggregate its composite rate for earth-station and space-segment services, which at the time was $1,060 per month for a voice-grade half-circuit. In response, Comsat set a monthly charge of $655 for the space segment and $405 for the ground segment, based on its estimate that space-segment costs accounted for about sixty-two percent of overall costs. The space-segment figure was reduced to $630 in October 1985 and to $610 in June 1986 (*see* Johnson 1987, p. 8). It will be recalled that the INTELSAT space-segment charge to Comsat and other signatories is presently $390 per month for a voice-grade half-circuit. Of the five international earth stations it owned as a result of this decision, Comsat retired two from service in 1986 and will sell the rest to AT&T in 1988 (Bruce, Cunard and Director 1986, p. 285).

Comsat has in general been ordered to conduct its newly authorized business activities, such as direct retailing of circuits and provision of earth-station facilities, through structurally and financially separate subsidiaries. The FCC, in its 1984 *Comsat Structure* order, allowed these subsidiaries to take advantage of certain legal, accounting, managerial and other support services of the »jurisdictional« parent corporation, while the latter continued to fulfil the mandate of the Communications Satellite Act (Sarreals 1986, pp. 288–89). A second major structural transformation of Comsat is presently under way in its ongoing merger with Contel, a major diversified U.S. telecommunications firm.

These developments were played out against the backdrop of an »appallingly long and complex« FCC rate proceeding against Comsat (Johnson 1987, p. 12) that lasted from 1965 until 1978. By 1975, when the Commission had reached its initial decision, the record had accumulated 20,000 pages (Johnson 1987, p. 13). The Commission determined that Comsat's cost of capital was 10.8

percent but allowed the corporation to earn up to 11.8 percent »if this can result from internal efficiencies and economies«; disallowed the inclusion of $152 million in Comsat's rate base that the corporation had capitalized to cover initial launch and satellite failures; and restructured Comsat's rates away from distance-sensitive, route-by-route tariffs to reflect distance-insensitive costs (Johnson 1987, pp. 13–14). The FCC's decision was stayed on appeal in 1975 and finally settled by an agreement of both parties in 1978, thirteen years after the rate case had begun. The agreement increased Comsat's allowed rate of return to 11.48 percent, again with a one percent bonus for efficiency; required a rate reduction from $2,850 to $1,575 per month so as to reduce overall revenues by 48.5 percent; and ordered Comsat to refund its carrier customers the approximately $100 million, held in escrow since 1975, that reflected the disputed tariff differential (Johnson 1987, p. 14–15).

2. *Comsat's present and future policy dilemmas and their effects on INTELSAT*

The most important challenge presently facing Comsat is the widespread sentiment in the U.S. government and in the business community that certain U.S. communications entities should be allowed »direct access« to INTELSAT, thereby circumventing Comsat and breaking its statutory monopoly as a wholesaler of INTELSAT circuits.

The FCC began an inquiry into direct access in 1982 but finally decided against it in 1984:

> [W]e find that were we to adopt direct access, we would merely be changing the form in which these expenses would be recovered, and in the process, adding an unnecessary layer of regulation with its attendant costs ... direct access of any form would not appreciably diminish any of the expense elements in the ... costs which comprise the space segment's portion of Comsat's tariff (*cited in* Johnson 1987, p. 28).

Johnson finds the FCC's reticence to allow direct access as being rooted in the »difficulty that neither the FCC, nor any other organization, has a benchmark for deciding the reasonableness of Comsat's costs or how these costs might be affected by direct access« (Johnson 1987, p. 28).

The executive branch, however, continues to look with favor on the arguments for direct access (*see* Bruce, Cunard and Director 1986, pp. 286–287). The authoritative interagency White Paper, for example, argues as follows:

> The process by which customers obtain international satellite communications services results in end-user prices substantially above INTELSAT circuit charges. The current arrangement where Comsat, in effect, functions as an exclusive U.S. marketing agent for INTELSAT circuits, may be ill-suited to an era of proliferating customer demands. No single entity, no mater how perceptive, can reasonably be expected to anticipate and satisfy all customer demands and needs in a market which is experiencing rapid demand-inducing and cost-reducing technological advances. Permitting expanded, direct, cost-based access to INTELSAT may be the most reliable means of substantially reducing costs and ensuring valid entry signals (U.S. Senior Interagency Group 1985, p. 32).

Most of this statement reflects the same predilections underlying the U.S. government's preference for systems separate from INTELSAT. Likewise, the National Telecommunications and Information Administration (NTIA) of the Department of Commerce has asserted that various benefits would flow from a policy of direct access. These include mitigation of marketplace distortions by removing policies favoring unequal facilities investment; downward pressure on space-segment prices due to the necessary unbundling of Comsat charges; discouragement of entry by »inefficient operators« due to the more precise alignment of costs and prices; greater use of the INTELSAT space segment due to competition by the greater number of firms accessing it; and transmission to all users of savings realized by large users accorded direct access (Bruce, Cunard and Director 1986, p. 287).

Not all observers favor direct access. Comsat, of course, is against it. Comsat's management contends that direct access would have a sharply adverse effect on INTELSAT. Multiple accessors in each country might dissipate the institutional and political coherence of the organization; in addition, there is no assurance that NTIA's claim of additional traffic stimulation by such competition would come to pass. The opposite, it is felt, might be the case.

Sarreals has stated the case against direct access as follows, drawing on certain implications of INTELSAT's cooperative financial structure:

> [T]he proponents of direct access believe that Comsat is nothing more than a sales agent for INTELSAT. In actuality, INTELSAT is the agent of Comsat and other Signatories, who are the investor/owners of the INTELSAT system. Unlike other international players, INTELSAT does not earn a profit ... and does not sell its services to an outside public. Rather, INTELSAT is a cost sharing *cooperative* which owns and operates satellites on behalf of its partner members, the Signatories. It is the Signatories, not INTELSAT, who put up the capital for the system, pay its operating expense, stand at risk for their investments, and market services directly to its customers ... INTELSAT has no commercial identity independent of the Signatories. Since INTELSAT does not exist apart from its members, the carriers' proposal to by-pass Comsat and obtain space segment directly from INTELSAT is without merit (Sarreals 1986, p. 285; emphasis original).

Another urgent challenge facing Comsat is the reform of its tariff structure to reflect route-by-route costs. We have seen, for example, that the outcome of the Comsat rate case in essence mandated continued dependence on average-cost pricing. As a wholesaler of INTELSAT's circuits, Comsat's need to compete with fiber-optic cables by pricing according to marginal rather than average cost is every bit as valid and urgent as that of INTELSAT itself. One rather radical solution to Comsat's pricing troubles is the replacement of rate-base regulation by a simpler criterion of price setting (Johnson 1987, pp. 50–51). The Office of Telecommunications (OFTEL) in the United Kingdom, for example, has experimented with an »RPI minus x« pricing rule for BT requiring that BT's price for each service rise no more rapidly each year than x per cent below the retail price index, where x is a small positive number such as three (*see* Jonscher 1986). Elimination of rate-base regulation for AT&T and Comsat would, among other things, do away with the Averch-Johnson effect, which is the bias toward overcapitalization induced by such a regulatory policy. The alleged presence of that bias, as we have seen, is the main justification for circuit-loading requirements balancing cable and satellite links (Johnson 1987, p. 51).

Comsat, if it continues to be constrained to average-cost prices, might fail to earn its authorized rate of return when excess capacity on fiber-optic cables begins to be marketed in 1988. In addition, Comsat's market power might be effectively removed by the advent of direct access and the diversion of traffic to cables. Under these circumstances, arguments would exist to deregulate Comsat entirely and perhaps allow it to launch and operate its own satellites in competition with INTELSAT.[16]

Although most or all of these policy alternatives for Comsat might have beneficial effects on U.S. consumers and other users, they would have a largely negative spillover on INTELSAT. Direct access, for example, would require the United States government to find some way for Comsat as signatory to consult with and give weight to the viewpoints of the other U.S. entities directly accessing INTELSAT. There is evidence that the United Kingdom's policy of allowing Cable & Wireless, which effectively enjoys direct access to INTELSAT, to have consulting rights with BT, the British signatory, has been successful. The effect of such power-sharing arrangements, however, cannot be seen as salutary for an organization already containing 113 members and very much in need of strong and unified leadership from its major users. The effect on INTELSAT of pricing

16 *See* Johnson (1987, p. 35). Johnson (1987, pp. 54–56) also supplies a catalogue of policy recommendations regarding Comsat.

innovations for Comsat and of its deregulation are more difficult to predict. The far-fetched option of Comsat establishing its own separate system in direct competition with INTELSAT, however, would place the corporation in a clear conflict of interest with its duties as an INTELSAT signatory. In such a case, the U.S. government might wish to appoint a different signatory.

In any event, Comsat's influence in INTELSAT has been massive over the past twenty-two years, although it has declined with the entry into force of the permanent agreements, the reduction in Comsat's share of INTELSAT's traffic and the expiration of Comsat's management contracts on behalf of INTELSAT. Nevertheless, deregulatory alterations of Comsat's structure, mission and tariff policy – both those already mandated and those that may yet occur – do not on balance appear to offer particular promise to INTELSAT, even though they may be beneficial to U.S. user and industrial interests.

D. *Deregulatory Trends in INTELSAT Member-States and Their Effects on INTELSAT*

Several industrialized members of INTELSAT – notably the United States, the United Kingdom and Japan – have taken major steps toward liberalizing and restructuring their domestic telecommunications industries during the past several years. Most others are considering steps in that direction. The first of the two subsections to follow will review the chronology of these developments and suggest their major causes. The second will examine the effects of telecommunications liberalization within individual INTELSAT members on the organization itself.

1. *Telecommunications liberalization in individual INTELSAT member-countries*[17]

Deregulation in the telecommunications sector of the United States has been under way since the 1960s. It was accelerated by an increasingly favorable

17 See Bruce, Cunard and Director (1986) and Snow (ed., 1986) for extensive studies of deregulatory trends in the telecommunications sectors of major industrialized countries. Subsection 1 draws extensively on these studies.

political climate that began with deregulation of the trucking and passenger-airline industries during the Carter administration. Competition in domestic satellite systems, for example, has existed since 1970. Competition with AT&T and other domestic and international common carriers has intensified during the 1980s with the advent of small, aggressive, specialized firms, such as MCI and GTE-Sprint. In 1982, long-standing federal litigation against AT&T was settled by a Modified Final Judgment, according to which the corporation was divested of its twenty-two regional common-carrier subsidiaries. These were reorganized into seven independent regional holding companies. Both these regional companies and AT&T itself were accorded broad new powers in new markets, both within and outside of traditional telecommunications areas. MCI, for example, has successfully concluded agreements with the Belgian PTT to carry traffic to and from that country. The U.S. situation is still fluid. There have been adverse political repercussions of the inevitable alignment of costs with prices brought about by competition after decades of subsidy patterns. Generally, rural users have been subsidized by urban users, and toll traffic has subsidized local traffic. The FCC, at the behest of Congress, has retarded the pace of implementation of a federal access charge levied on all telephone subscribers to help approximate prices to underlying costs.

Liberalization in the United Kingdom occurred primarily as a result of the Thatcher government's desire to stimulate competition in the British economy and to reduce the entrepreneurial role of government. By 1985 British Telecom (BT), the major domestic and international carrier, had been privatized by an immense flotation of shares to the public. In addition, the Mercury consortium, with major financial backing from Cable & Wireless, was licensed to compete with BT over a broad range of markets. BT received a new operating license, and the Office of Telecommunications (OFTEL) was established to regulate these newly realigned entities according to the applicable legislative and regulatory guidelines. Partly as a result of trade-union opposition to interconnecting Mercury with BT's network to allow it to compete, Mercury to date has made only minor inroads into BT's markets. BT, however, stimulated at least in part by the threat of Mercury, has become boldly entrepreneurial in markets throughout the world and has shown greater responsiveness in catering to user needs for efficiency, low prices and diversity of new services.

Similar legislation occurred in Japan at about the same time as in the United Kingdom. NTT, the domestic carrier, was privatized, and a second piece of legislation admitted certain kinds of competition. KDD, the international carrier that also serves as Japan's INTELSAT signatory, has been in private hands since its formation in the 1950s but has not yet been subjected to competition. The Japanese strategy for liberalization was more pragmatic and less ideological than that motivating broadly comparable initiatives in the United Kingdom. Japan's Ministry of International Trade and Industry (MITI) saw privatization and competition as part of an overall »industrial strategy« to strengthen the Japanese economy sector by sector and to preserve and enhance its international competitiveness. The Ministry of Posts and Telecommunications (MPT) did not fully subscribe to this approach, but a consensus was eventually arrived at that preserved many of MPT's institutional and regulatory prerogatives.

Continental European countries have not been similarly active in liberalizing their telecommunications sectors. As we have noted, the French maintain that a cooperative approach holds many of the same advantages of competition without the latter's alleged tendencies toward proliferation of differing technical standards, dissipation of scale and scope economies, and encouragement of uneconomic entry. The numerous telecommunications subsidiaries of the French PTT, to be sure, are vigorously competitive in markets throughout the world, including the United States. Recent electoral reversals of President Mitterrand's party, as well as the approach of the end of his term of office, have somewhat attenuated the French stance against liberalization.

In West Germany, a government commission is preparing a detailed study of telecommunications policy that will be released in late 1987. Many other similar studies have preceded it, the majority of which have been rather favorable toward various forms of liberalization. There is however little prospect that the Bundespost will adopt major reforms, like privatization or widespread competition in network facilities. Such developments appear to be precluded, among other things, by the ongoing constitutional and jurisdictional conflicts between the states (*Länder*) and the federal government in telecommunications and broadcasting matters.[18] Another obstacle is posed by the uneasy political consensus regarding the appropriateness of subsidies inherent in Bundespost tariffs to achieve various social objectives.

18 The uneasy truce reached between the federal government and the *Länder* with respect to Bildschirmtext, the West German videotex system, is illustrative in this regard. *See* Bruce, Cunard and Director (1986, pp. 576–80). Any analysis of broadcasting and telecommunications in West Germany must take into account the fact that the Basic Law grants the *Länder* the fundamental ownership, operational and oversight responsibilities in broadcasting, while the federal government, through the Bundespost, is in charge of telecommunications. The difficulty of institutional change in these areas, however, is more a function of political opposition to such change – primarily through a coalition of trade union support and major Bundespost suppliers – than of this federal-state dichotomy itself.

Italy, Belgium, the Netherlands and Luxembourg are examining various alternatives that would introduce a greater measure of competition and privatization into their telecommunications sectors. Australia and New Zealand, by contrast, do not appear to favor any such moves in the near future. Canada, like West Germany, faces urgent legal and constitutional conflicts between its provincial governments and the national government in Ottawa that must be clarified before decisive action on further liberalization can be taken. Canada's proximity to the United States and its extensive symbiotic interchange with the U.S. economy, however, have already brought about a certain level of private ownership and competition in telecommunications.

There have been various proximate causes of liberalization, as we have seen, in each country. In the United States, it was a long-term, institutionally mature evolution toward deregulation of much of the economy. In the United Kingdom, developments were more dramatic and rapid and had a greater basis in the political philosophy of the present government. Events in Japan were well-considered pragmatic responses to challenges of the economy's industrial structure and long-term international competitiveness. Yet underlying all of these actions and immediate motivations was the pervasive dynamic of improvements in telecommunications technology, a force we have noted previously. These changes transformed the cost functions underlying all output techniques and thereby implied changes in the legislative, regulatory and technical conduct of telecommunications. What effects have these changes in individual INTELSAT members had on the organization itself?

2. *Consequences of domestic liberalization on INTELSAT as an organization*

By far the greatest fallout of domestic telecommunications restructuring on INTELSAT has occurred as a result of events in the United States. The perceived economic success of domestic deregulation in trucking, aviation, telecommunications and other sectors has led the executive branch in particular to imagine that similar successes could be realized with comparable ease in the international arena. Often such thought has rather naively discounted the difficulty of convincing foreign administrations exhibiting quite diverse political, regulatory and industrial conditions to engage in domestic and international deregulation. Frustration at the slow pace of cooperation with foreign partners in bringing about international telecommunications deregulation may well have contributed to the aggressiveness with which the

U.S. government has supported the entry of separate international satellite systems into competition with INTELSAT.

More subtle effects may be in the process of attenuating the unconditional support of INTELSAT by various of its other member-countries. British Telecom, as we have already noted, must now consult with Cable & Wireless regarding U.K. policy toward INTELSAT, since Cable & Wireless enjoys access to INTELSAT's space segment. Certainly consultation arrangements such as this, which might also be in the offing for the U.S. signatory if entities other than Comsat acquire direct access to INTELSAT, will weaken the force and authority with which INTELSAT's constituent parts speak to its director general and executive organ.[19] Such an organization would certainly be less able aggressively to pursue its interests by competing with rival satellite or fiber-optic cable systems.

INTELSAT's signatories assume differing roles in their own countries. Some, as we have noted, already confront entry pressures from newly authorized domestic carriers, and others may soon face such competition. This may induce the signatories to price INTELSAT's space segment at marginal cost in their domestic markets. Such a policy, in turn, might make inroads by fiber-optic cables more difficult than they would be if the average-cost pricing typical of regulated monopolists were perpetuated. At the retail level, this would mean that INTELSAT's end users would find greater price responsiveness and a broader variety of services. On the other hand, these very competitive pressures would cause INTELSAT's signatories to demand more efficiency and pricing discipline from INTELSAT itself. This would in turn contribute to the set of circumstances already constraining INTELSAT to become more »corporate,« flexible in its pricing and perhaps less able to subsidize its developing members both explicitly and by means of the redistribution of revenues implicit in its average-cost tariff policy.

19 The analogy of a price cartel suggests itself. Such an arrangement provides a maximum of stability and joint benefits if no one violates the underlying rules. Even greater benefits, however, accrue to individual cartel members that can cheat and escape detection and sanctions by the others. This incentive to cheat makes cartels inherently unstable organizations, as the history of the Organization of Petroleum Exporting Countries (OPEC) or the International Civil Aviation Organization (ICAO) indicates.

E. *Effects of Limited Launch Capacity*

Launch failures in 1986 by the vehicles anticipated to lift off the INTELSAT VI series of satellites – NASA's space shuttle[20] and the European Space Agency's Ariane – underscored the fragility of the link from the earth to geostationary orbit. The resulting uncertainty in the availability of launch services from the world's two largest and hitherto most dependable suppliers has had two effects on INTELSAT. First, it has enhanced the competitive threat of fiber-optic cables relative to that of private separate satellite systems: none of the latter are yet launched. Second, it has caused INTELSAT and other satellite operators to look beyond NASA and the ESC to other sources of launch services.

It is clear that the space shuttle and Ariane failures have had a chilling effect on the many separate systems that wish to compete with INTELSAT. PanAmSat, having completed a successful Article XIV(d) coordination procedure with the Assembly of Parties, anticipates a mid-1987 Ariane launch date. Other systems, none of which have yet secured foreign partners or completed Article XIV(d) procedures, are in a much more difficult position. Launch availability alone is a major reason why there will be no plethora of separate systems following the presumably successful commencement of operation by PanAmSat.

INTELSAT, of course, faces the same launch uncertainties as its would-be competitors, but its long record of operation and its status as an intergovernmental cooperative with a broad membership base might well give it the political influence and institutional experience it needs to gain priority access to such launch dates as now exist. In investigating other sources of launch services, however, INTELSAT is merely attempting to diversify the considerable risks in launch availability that now exist.

One possibility is the Proton series of launchers offered by the Soviet Union. INTELSAT anticipates various U.S. objections to the use of Soviet launchers, however. U.S. munitions-control regulations have long posed formidable obstacles to the export of U.S.-manufactured hardware such as satellites with

20 Several reliable references on the U.S. space program are available. The study of the U.S. Office of Technology Assessment (1984) is a comprehensive treatment stressing international cooperation and competition. The U.S. National Commission on Space (1986) issued its report shortly after the space shuttle accident. The U.S. Presidential Commission on the Space Shuttle Challenger Accident (1986) is a detailed report of the January 1986 tragedy. The report of the Business-Higher Education Forum (1986) is a representative statement of academic and industry views of the future of the U.S. space program.

even a remote potential for military application and divulgence of sensitive technology to unfriendly countries. Although an export license was issued for INTELSAT V spacecraft launched by Ariane from the French spaceport at Kourou in French Guiana, it would be much more difficult to convince U.S. authorities to allow the export of a U.S.-built satellite to the Soviet Union for launch. Extraordinary measures could presumably be taken to justify an export license. These could include, for example, keeping the spacecraft in a sealed container prior to its insertion into the Proton's payload and allowing U.S. or third-country observers to accompany and monitor the spacecraft continuously during the period between its export and launch. A final commercial possibility is at present the Long March rocket of the People's Republic of China. In addition, Japan may well offer commercial launch services into the geostationary orbit before long.

F. *INTELSAT as an International Organization*

This section examines INTELSAT's status und relations among the community of international organizations. During its interim period, INTELSAT maintained an exceedingly low profile as an international organization. Dependent on U.S. technology, management and usage, and subject to Comsat's absolute veto power in the ICSC, INTELSAT rented rooms in Comsat's Washington office headquarters for several years and reportedly had no entry in the telephone directory for much of that time.

This has all changed with the advent of the permanent agreements. INTELSAT now occupies its imposing new headquarters building in Washington; has an executive organ of some six hundred full-time international civil servants[21] enjoying salary levels and conditions of employment comparable to those in other international organizations; is governed by permanent multilateral treaties according both its large and small users well-balanced powers of access and influence; and presides over a technology that has rapidly spread among a large number of industrialized countries, allowing it broad scope in choosing among launch and spacecraft contractors.

Subsection 1 outlines the establishment of INTELSAT's privileges, exemptions and immunities in accordance with Article XV of the intergovernmental

21 As of February 1986, INTELSAT had a staff of 606 and an authorized staffing level of 651. *See* MS-16-10E, March 12, 1986, p. 12.

agreement, while Subsection 2 discusses its formal and informal relationships with other international organizations.

1. *INTELSAT's privileges, exemptions and immunities*

Article XV(b) of the intergovernmental agreement provides that INTELSAT members exempt INTELSAT and its property from national income and property taxation and from customs duties. Article XV(c) requires the United States, as the »Party in whose territory the headquarters of INTELSAT is located,« to

> conclude a Headquarters Agreement with INTELSAT covering privileges, exemptions and immunities. The Headquarters Agreement shall include a provision that all Signatories in their capacity as such, except the Signatory designated by the Party in whose territory the headquarters is located [i.e. Comsat], shall be exempt from national taxation on income earned from INTELSAT in the territory of such Party [i.e. the United States government]. The other Parties shall also as soon as possible conclude a Protocol covering privileges, exemptions and immunities.

A Headquarters Agreement was signed by INTELSAT's secretary general on November 22, 1976 and by the U.S. Department of State on November 24, 1976. On the basis of an executive order of the U.S. president on January 19, 1977 and an exchange of letters between the director general and the Department of State on February 1, 1977, the Headquarters Agreement entered into force retroactively as of November 24, 1976 (MS-6-10E, April 1977, p. 10).

A draft Protocol on INTELSAT Privileges, Exemptions and Immunities was circulated among INTELSAT members in April 1977; revised according to comments received; and recirculated in March 1978. From May 8 through 19, 1978, INTELSAT members attended a conference in Washington to draft a Protocol for presentation to INTELSAT's member-countries (MS-7-10E, March 29, 1978, p. 9). The draft provided that it would enter into force on the thirtieth day after the date of deposit of the twelfth »instrument of ratification, acceptance, approval, or accession.« It was transmitted by the director general to all INTELSAT members in June 1978 (MS-8-10E, March 14, 1979, p. 6).

Almost two years later, as of March 10, 1980, only nine INTELSAT members had completed the ratification process for the Protocol. The Board of Governors admonished INTELSAT's other members that it would be in the organization's interests »if the necessary actions by the Parties ... could be accelerated« (MS-9-10E, March 18, 1980, p. 6). The twelfth instrument of ratification was finally deposited with the director general on September 9, 1980, and the Protocol duly entered into force one month later (MS-11-10E, March 20, 1981, p. 7). Whether through inattention or because of objections on the part of INTELSAT members to portions of the Protocol, only 26 parties, or less than one-fourth of INTELSAT's membership, had formally ratified it as of March 1986 (MS-16-10E, March 12, 1986, p. 10).

2. *INTELSAT's relationships with other international organizations*

a. *Relationships with the ITU*

It is the ITU with which INTELSAT has maintained the closest and most sustained relationship over the years. Withers and Weiss (1984) provide an exhaustive compendium of the technical and administrative aspects of that relationship, including information regarding the evolving role of the ITU and its regulations in the Article XIV coordination procedure. What has been the nature of the INTELSAT/ITU relationship since the mid-1970s?
Throughout its existence, INTELSAT has sent observers in an advisory and consultative capacity to various ITU meetings, including study groups and plenary assemblies of its consultative committees, world and regional administrative radio conferences, regional plan committees, symposia and exhibitions. Yet INTELSAT itself, sincc it is not the government of a national state, cannot join the ITU. During the latter part of the 1970s, exploratory discussions were held between representatives of the two organizations regarding the establishment of formal relations. Early in 1980, the ITU's secretary general wrote to INTELSAT stating his view that »working relationships between the two Organizations were satisfactory, expressing his willingness to enter into an agreed arrangement ..., and stating that any additional agreement ... would require the agreement of the members of the ITU at a Plenipotentiary Conference« (MS-9-10E, March 18, 1980, p. 9).

In July 1980, an agreement on »Arrangements between the Secretary General of the ITU and the Director General of INTELSAT« was adopted. It provides for the exchange of documents and publications; the exchange of information regarding conferences and meetings; and »consultation concerning questions deemed to be of common interest for the development of telecommunications« (MS-11-10E, March 20, 1981, pp. 9–10). These rather informal arrangements appear to have suited the needs of the two organizations, as the

question of a formal institutional relationship between them has not been pursued since that time.
INTELSAT's most important challenge affecting its ITU relationship came during the 1985 Space WARC meeting, at which time INTELSAT attempted to convince the conference delegates of its need for priority assignments of orbit and spectrum as a »multi-administration system created by inter-government agreement and used collectively.« The ITU made no specific or binding provisions to that effect; instead, it agreed simply to »take into account« the »requirements« and »specific characteristics« of organizations such as INTELSAT (MS-16-10E, March 12, 1986, pp. 14–15).

b. *Relationships with international organizations other than the ITU*

Article XXI of the intergovernmental agreement obliges INTELSAT to submit an annual report to the secretary general of the United Nations. In addition, INTELSAT representatives have attended various United Nations meetings in a more or less official capacity. For example, INTELSAT was represented at sessions of the United Nations Committee for the Peaceful Uses of Outer Space (COPUOS) held in 1984 and 1985 and at a COPUOS committee and subcommittee in 1985 (MS-15-10E, April 2, 1985, p. 12; MS-16-10E, , March 12, 1986, p. 14). INTELSAT observers were also present at the Second United Nations Conference on the Exploration and Peaceful Uses of Outer Space (UNISPACE) in 1982 (MS-13-10E, March 24, 1983, p. 12). Finally, in 1984 the United Nations and INTELSAT signed an agreement providing for satellite capacity from INTELSAT for U.N. peacekeeping and emergency relief activities. The agreement is for the five-year period commencing September 8, 1984 (MS-15-10E, April 2, 1985, p. 13).
Since INMARSAT was created along the lines of INTELSAT and has leased INTELSAT capacity to help provide services to its members, it is not surprising that the question of formal relationships between the two organizations has arisen. INTELSAT representatives attended early meetings of INMARSAT's Preparatory Council, Technical and Operational Advisory Committee, and Council (MS-9-10E, March 18, 1980, p. 11). An agreement establishing relations between the two organizations was approved by the Assembly of Parties late in 1982 (MS-13-10E, March 24, 1983, p. 13). Another ocean-related entity, the International Maritime Consultative Organization (IMCO), established formal relations with INTELSAT by an agreement ratified by the Assembly of Parties in 1976 (MS-6-10E, April 1977, p. 9).
Four memoranda of understanding (MOUs) establishing relations between INTELSAT and other organizations are noteworthy. The first MOU was

signed with the Asia Pacific Telecommunity, a regional intergovernmental organization under United Nations auspices with headquarters in Bangkok. Its objectives include planning, development and coordination of technical standards and network routing plans (MS-11-10E, March 20, 1981, p. 11). The MOU was approved by the Assembly of Parties late in 1982 (MS-13-10E, March 24, 1983, p. 13). Three other MOUs were signed in 1985. The Assembly of Parties ratified such pacts with EUTELSAT and with the Intergovernmental Bureau for Informatics (IBI) im Rome. In addition the Assembly »acknowledged« an MOU between INTELSAT and the Ministry of Posts and Telecommunications of the Soviet Union (MS-16-10E, March 12, 1986, p. 15).
Finally, INTELSAT has become a member of the Pacific Telecommunications Council (PTC), a non-governmental, non-profit organization with headquarters in Honolulu that sponsors a well-attended conference there early each year. This has proven to be a convenient forum for the informal exchange of views and information regarding Pacific telecommunications among government representatives, business people, academics and members of international organizations. INTELSAT representatives have attended these meetings regularly, served on the PTC Board of Trustees and presented papers about INTELSAT's activities in the Pacific rim and basin and on behalf of developing countries generally.

G. *INTELSAT and the Developing Countries*

INTELSAT's less-developed members have come increasingly to regard the organization as their advocate. They correctly perceive that their interests are threatened by the technological, economic and political pressures that are forcing INTELSAT to change and adapt in order to survive such threats as separate satellite systems and fiber-optic cables. This section examines the relationship between INTELSAT and the developing world. Subsection 1 reviews the literature establishing linkages between telecommunications infrastructure and economic development. Subsection 2 notes the tariffs and earth-station standards that INTELSAT has established with the needs of Third World countries in mind. Subsection 3 describes the major demonstration, aid and training programs that INTELSAT has developed and implemented to satisfy the unique needs of developing countries. Finally, Subsection 4 contains several preliminary conclusions.

1. *The role of telecommunications in economic development*

Recent research has underscored the vital role of telecommunications infrastructure in providing a basis for ongoing economic development.[22] Many of the early studies, to be sure, were merely crude correlations of telephones per hundred with gross domestic product per capita. Others betrayed considerable econometric naivete by focusing too exclusively on the direction of causality between telecommunications and development. Today, econometric models using simultaneous equations can specify causal patterns in both directions and relate them to other macroeconomic forces at the same time.

Several studies are of particular value. Saunders (1983) adds to the fund of information and institutional knowledge about political and institutional barriers to establishing modern telecommunications in the Third World. Hudson (1983) synthesizes the findings of ITU and OECD studies during the early 1980s that related telecommunications to economic development. In another book (Hudson 1984), she examines methodological and empirical preconditions for understanding the effects of telecommunications in rural areas. The volume by Sanders, Warford, and Wellenius (1983), who are World Bank economists, distills rich institutional detail from their broad experience in telecommunications development projects. They argue that cost-benefit analysis does not capture the full contribution of telecommunications to development because of the pervasive spillover effects (externalities) that are not captured by the pricing system or by cost-benefit methodology. Leff's pieces (1984a, 1984b) provide a theoretical basis for discovering and quantifying the role and extent of externalities in telecommunications projects. Pierce and Jequier (1983) present the results of collaborative research by the ITU and the OECD. Among other things, these studies estimate the economic benefits that a »global developmental satellite system« would generate. Maitland (1984), who directed an ITU-mandated inquiry into telecommunications and development, underscores the importance of the telephone in the development process. His commission's report singles out inadequacies in organization, management and training as particular barriers on the supply side. A recent series of essays sponsored by the ITU (1986) contributes to the goal of more exactly quantifying the benefits of telecommunications. Gille (1984, 1986), one of the contributors, posits a multisectoral model using elements of neoclassical equilibrium and growth theory while capturing many of the particular features of telecommunications markets.

22 Hudson *et al.* (1979) and Snow and Jussawalla (1986, Chapter 14, pp. 169–78) provide annotated bibliographies of this topic.

Elsewhere, the ITU studies conclude that all economic strata of the population, and not just the urban elite, benefit from telecommunications availability; that businesses as well as government and consumers can save money through improved telecommunications; and that public call boxes in rural areas are particularly beneficial. Snow (1985b), using the Pacific rim and basin as an example, argues that liberalization of the telecommunications sector not only accompanies the process of economic development but appears to encourage it as well. Pelton (1982) and Rahim and Wedemeyer (1983) also use data from the Pacific to document the link between telecommunications and development.

Several studies explicitly link satellite communications to economic development. Recent contributions by Hudson (1985) and Pelton and Demac (1986) are among the best. INTELSAT (INTELSAT 1986a) has provided a pamphlet outlining its own development-oriented policies and programs and summarizing information about other international organizations that are active in communications development. The particular strengths of satellite systems in the development process cluster around their ability to perform three primary functions: to link the LDC's capital city and its emerging financial markets to the rest of the world so as to facilitate trade and information exchange; to link rural areas to the capital so that the market economy and instantaneous financial information can penetrate the hinterland; and to provide infrastructure for the delivery of developmental programs in vital areas such as health, education and agriculture. The latter two functions, which extend telecommunications into areas with typically thin or often non-existent terrestrial telecommunications infrastructures, are particularly dependent for their financial viability on small, inexpensive earth stations and tariffs favorable to such terminals.

2. *INTELSAT tariffs and earth-station standards favorable to developing areas*

For broad coverage to be realized, earth stations must be relatively numerous in rural developing areas having little or no access to reliable terrestrial networks. For this and many other reasons, terminals must be small and therefore inexpensive so that they do not overly strain the scarce foreign-exchange reserves and communications budgets of developing countries. Originally, INTELSAT's earth-station standards, reflecting the technology of the 1960s and INTELSAT's membership of primarily industrialized countries, dictated thirty-meter dishes that cost several million dollars each and were

Table 3 – INTELSAT Earth Station Standards

Earth Station Standard	Antenna Size In Meters	Type of Service	Frequency Band (GHz)
A (to 1986)	30-32	International voice, data, IBS	6/4
A (1986 and beyond)	15-17	International voice, data, TV, IBS	6/4
B	10-13	International voice, data, TV, IBS	6/4
C (existing)	15-18	International voice, data, TV, IBS	14/11
C (revised)	11-13	International voice, data, TV, IBS	14/11
D1	5	VISTA (international or domestic)	6/4
D2	11	VISTA (international or domestic)	6/4
E1	3.5-4.5	IBS (K-band)	14/11 & 14/12
E2	5.5	IBS (K-band)	14/11 & 14/12
E3	8-10	International voice, data, TV, IBS (K-band)	14/11 & 14/12
F1	4.5-5	IBS (C-band)	6/4
F2	7.5-8	IBS (C-band)	6/4
F3	9-10	International voice, data, TV, IBS (C-band)	6/4
G	0.8-12	INTELNET, international TV, etc.	6/4 & 14/11
Z	4.5-18.5	Domestic voice, TV, data	6/4 & 14/11

Source: INTELSAT Report 1985-86, p. 18

designed for interconnection with dense terrestrial networks at only a small number of gateway locations. Since then, however, increases in launcher payload capacity and satellite power, as well as the consistent advocacy of INTELSAT's growing number of member-countries, have led INTELSAT to adopt earth-station standards and associated rate adjustment factors that make the use of small terminals less of a financial burden on developing countries. Table 3 indicates the present authorized sizes, service applications and frequency bands of earth stations authorized to access INTELSAT's space segment. It should be reiterated that earth stations and terrestrial networks are not part of INTELSAT but belong instead to its individual signatory organizations. As has been noted, earth-station and other terrestrial costs

Table 4 – Use by Non-standard Earth Stations in General

(a) Satellite utilization through non-standard earth stations leased on a unit basis is charged for in multiples of appropriate rates applicable to similar use through standard earth stations. The appropriate rate adjustment factor to be applied will be determined on a case-by-case basis taking into account the particulars of the earth station and the actual services to be provided.

(b) The established rate adjustment factors related to operations in the 4 GHz downlink are as follows:

Station* G/T (dB/K)	Rate Adjustment Factor
35.0 or greater (with in increase in satellite power required)	1.0
35.0 or greater (with an increase in satellite power required)	2.5
31.7 - 34.9	2.5
28.7 - 31.6	5.0
25.7 - 28.6	10.0

(c) The established rate adjustment factors related to the operations in the 11 GHz downlink are as follows:

Station* G/T (dB/K)	Rate Adjustment Factor
33.0 or greater (with no increase in satellite power required)	1.0
33.0 or greater (with an increase in satellite power required	2.5
28.7 - 32.9	2.5
25.7 - 28.6	5.0
22.7 - 25.6	10.0

* For stations in a defined network consisting of two or more stations wherein non-standard earth stations are approved for limited access to any operational satellite in the global INTELSAT system.

Source: INTELSAT Tariff Manual, S-86-22, June 17, 1986, pp. 56-57

amount to between eighty and ninety percent of most end-to-end connections using satellite links.

INTELSAT's Tariff Manual (S-86-22, June 17, 1986) establishes rate adjustment factors and other terms and conditions of use associated with »non-standard« earth stations (i.e., earth stations other than those of standard type A as shown in Table 3). A number of these were noted in the discussions of INTELSAT's tariff policy in Section A of Chapter III. The rate structure is adapted to smaller terminals on the basis of a technical criterion, namely the gain-to-noise temperature ratio (G/T), a measure correlating roughly with the diameter and cost of the terminal. For example, standard B stations using INTELSAT IV and IV-A satellites for voice traffic and having a G/T of 31.7 dB/k are charged 1.5 times the rate for standard A stations (INTELSAT Tariff Manual, S-86-22, June 17, 1986, p. 55). Table 4 indicates rate adjustment factors for non-standard earth stations in general, which depend on the downlink frequency, the G/T ratio and whether an increase in satellite power is required.

Other aspects of INTELSAT's tariff structure are also of particular interest to developing counties. Since 1973 INTELSAT has offered long-term leases of whole, half and quarter transponders to its members on an interruptible basis for domestic use. Over thirty countries have made use of this arrangement; most of them have been from the developing world. SPADE, a long-standing demand-assignment technique that is attractive for use on thin traffic links, is presently tariffed at $0.055 per minute of satellite holding time at each end of the circuit (INTELSAT Tariff Manual, S-86-22, June 17, 1986, p. 12). We shall now review particular INTELSAT policies, deriving both from its tariff and from specially mandated programs, that have been of special relevance to the developing world.

3. *INTELSAT's demonstration, aid and training programs*

a. *VISTA low-density telephone service* (*see* INTELSAT 1986a, pp. 16–24)

VISTA service, introduced in 1983, and SUPER VISTA, begun in 1986, were made possible by INTELSAT's relaxation of certain mandatory earth-station performance standards. These services use standard D earth stations (*see* Table 3). In addition, INTELSAT acts as a purchasing agent on behalf of VISTA users to keep costs down. Terminals cost as little as $70,000. SUPER VISTA addresses the burden of high space-segment charges by using demand-assigned multiple access (DAMA), a technique allowing more channels to be fit into a single transponder. Both VISTA and SUPER VISTA can be used for domestic or international connections on a preemptible or non-preemptible basis. Through its INTELSAT Assistance and Development Program (IADP), to be described below, INTELSAT offers aid in network design and implementation planning to VISTA users and user groups. Possible uses of VISTA and SUPER VISTA services include domestic and regional economic development projects; networks for regional and international organizations working in developing areas; and airport and air-traffic control networks.

b. *INTELNET services* (*see* INTELSAT 1986a, p. 25)

INTELNET networks address the needs of government and private agencies to gather and distribute information to remote locations practically and inexpensively. INTELNET terminals, corresponding to standard G performance characteristics, are easily transportable and can be as small as 65 cm in diameter by using spread-spectrum digital modulation techniques. For example, weather information, news, commodity prices or agricultural extension services can be sent or received at the presently approved data transmission rate of 9.6 kbps. Non-preemptible INTELNET service is tariffed at rates consistent with those for full and fractional INTELSAT Business Service (IBS) transponders. Preemptible INTELNET service on spare capacity is tariffed at rates consistent with preemptible transponder video leases. INTELNET service was developed through cooperation among INTELSAT, the Intergovernmental Bureau for Informatics (IBI) and Telespazio, INTELSAT's Italian signatory.

c. *Project SHARE* (*see* INTELSAT 1986a, pp. 27–33; INTELSAT 1986b)

Project SHARE (Satellites for Health and Rural Education) is an extraordinarily successful series of demonstrations involving the ability of satellite technology to aid in the process of development. It was inaugurated on January 1, 1985 and, after two extensions, will cease operation at the end of 1987. In collaboration with the International Institute of Communications (IIC) in London and SHARE's International Advisory Council, INTELSAT has provided free use of its satellites for tests and demonstrations in the fields of health care and education. Initial scepticism by the Board of Governors was overcome by the enthusiastic response of participants from forty-four countries in some twenty projects. The most ambitious project was a tele-education demonstration in the People's Republic of China. An initial phase involved nationwide video broadcasts of advanced academic courses six hours a day to forty-five locations. The follow-on phase expanded the original network to fifty-three locations and included more diverse material, such as programs on health, calligraphy and fine arts. Following the project, the Chinese government leased two transponders from INTELSAT to continue similar programs.

d. *IADP and IDF* (*see* INTELSAT 1986a, pp. 10–13)

The INTELSAT Assistance and Development Program (IADP), begun in 1978, provides operational and technical help to INTELSAT's signatories and users. This assistance, which has been extended to eighty-nine countries, is generally furnished free of charge. It has included seminars relating to earth-station technology and new INTELSAT services. Applications include advice to countries (which are typically not yet INTELSAT members) planning their first earth station; countries introducing new earth-station techniques; and countries planning a domestic leased satellite system. Future IADP emphasis is likely in the areas of training programs, reassignment of surplus earth-station equipment and advice on financing.
The INTELSAT Development Fund (IDF), established in March 1985, addresses the needs of developing countries not able to make full use of INTELSAT's existing developmental programs, such as VISTA, INTELNET and Project SHARE, because of lack of funding. The IDF is not a permanent source of funding; rather, it provides assistance to countries in locating sources of external telecommunications aid or funding and in defining their initial needs both for satellite circuits from INTELSAT and for integrating their domestic and international services. Projects are considered on a case-by-case basis.

e. *Programs for visitors at INTELSAT headquarters* (*see* INTELSAT 1986a, pp. 44–46)

Three INTELSAT programs bring young people and mid-career professionals to INTELSAT headquarters to acquire job skills and scholarly insights having to do with satellite communications. Under the aegis of the INTELSAT Assignee Program, which was created in 1965, signatories nominate personnel for one-year terms (renewable for a second year) involving technical interaction with members of INTELSAT's executive organ. To date 108 assignees from twenty-five countries have participated. About half the countries are from developing areas.

The INTELSAT Young Professionals Program, inaugurated in 1982, allows ten persons nominated by signatories to gain skills and training during a year at INTELSAT headquarters that will be of help upon their return home. Participants must be between twenty-two and thirty-three years of age. A specific curriculum is designed to meet the requirements of each participant. To date, fifteen young professionals representing thirteen developing countries have taken part in this program.

Finally, the INTELSAT Visiting Scholar Program, launched early in 1986, allows graduate students from INTELSAT signatories to gain work experience at INTELSAT. Stress is laid on training in the relevant branches of science and engineering, and supplemental study at Washington-area universities is permitted. The first three participants are from Bolivia, the People's Republic of China and the Republic of Korea.

4. *Conclusions*

Scholarly work and field experience to date leaves no doubt that telecommunications infrastructure and economic development are closely intertwined. In addition, anecdotal evidence and common sense suggest that a modicum of communications links is necessary for a market economy to have any hope of establishing itself. Satellites, of course, play a vital role in providing such infrastructure.

While INTELSAT originally satisfied primarily the needs of its industrialized members, it has in recent years paid increasing attention to the requirements of the developing world. This can be attributed to some combination of benevolence, self-interest (the threat of competition from separate systems) and politics (the increased influence of small users under the permanent agreements). Whatever its sources, the evidence of INTELSAT's increased concern and support for the Third World is undeniable.

This gradual but massive reorientation has not been without its strains within the organization. The Board of Governors, for example, was, in INTELSAT's own words, initially »dubious about the value of Project SHARE« (INTELSAT 1986b, p. 2). By contrast, many developing member-countries felt that INTELSAT's efforts in the development area were inadequate. Indeed, pressure from the Third World within the Board of Governors, Meeting of Signatories and Assembly of Parties was instrumental in establishing the IADP. An early proposal to the Assembly of Parties by the Ivory Coast for an »INTELSAT Solidarity Fund,« for example, was found by the Board to be »inconsistent with the provisions of the INTELSAT Agreement and the

Operating Agreement and absent modification of these Agreements could not be implemented« (MS-7-10E, March 29, 1978, p. 10). A group of Arab countries on the Board subsequently submitted a proposal urging the examination of »Marketing Activities of INTELSAT to Cover the Needs of Developing Countries« (MS-7-10E, March 29, 1978, p. 10). It was the proposal of the Arab countries that led to the establishment of the IADP the following year (MS-8-10E, March 14, 1979, p. 11).

Two conclusions suggest themselves. First, if INTELSAT's accelerated concern with developing countries is largely a reaction to the probability of entry by separate systems, this might well imply that the owners of such systems ultimately wish to serve developing countries as well as the dense traffic streams between industrialized countries. Otherwise, why would INTELSAT take such extensive measures to service and protect its markets in LDCs? It will be recalled, in this regard, that Peru, rather than an industrialized nation, is the only country to date to declare itself a partner of any of the various private U.S. systems. Second, the inevitable realignment of route-by-route costs with route-by-route prices that INTELSAT must undergo to meet the competitive challenge of separate systems will have a severely damaging effect on its ability to maintain the developmental services and programs of the kind outlined in Subsection 3 above. Such activities are clearly subsidized by the revenues that INTELSAT earns from lucrative routes on which it presently faces no satellite competition. Even without separate-system entry for the foreseeable future, competition from fiber-optic cables might be sufficient to attenuate INTELSAT's support of developmental aid programs. Although economic theory implies that direct subsidies such as grants are a more efficient means of redistribution than are indirect subsidies implicit in the pricing mechanism, it is not at all clear which international institutions, if any, would be prepared to furnish direct cash grants for telecommunications infrastructure at such time as INTELSAT is no longer politically or financially able to support its own indirectly funded efforts. As we have noted, many of the benefits of telecommunications infrastructure cannot be accurately measured, while its costs are manifest. Thus, present benefit-cost ratios are artificially low and make only an extremely conservative case for telecommunications aid. The loss of INTELSAT's increasingly substantial support to Third World telecommunications development, if and when it occurs, would be a severe blow that must be redressed at the highest levels of international policy making.

Chapter V: Conclusions

Although we have ventured judgments throughout this study, Chapter V is devoted to a more systematic attempt at drawing conclusions about INTELSAT from the material and discussions that have been presented in the preceding chapters. Section A posits five conceptual frameworks within which INTELSAT can be evaluated. The first – functionalist theory – has already been briefly mentioned in Section B of Chapter III, which reviewed INTELSAT's contributions to the peaceful use of outer space. The second – cooperation and social economy – was noted in Section A(3)(a) of Chapter IV, which summarized arguments of the West German *Gemeinwirtschaftslehre* in favor of monopoly provision of vital public services. Political economy of deregulation, an important new school of thought in regulatory economics, has not yet been mentioned. The collective-goods model, the fourth framework of analysis, was adumbrated in Section C(1) of Chapter III, which discussed its application to principles for allocating the spectrum and the geostationary orbit. The fifth and final framework of evaluation to be presented is that of neoclassical economics. This context has been implicit in much of the economic reasoning of the preceding chapters, but its premises and implications will be made more explicit in Section A of the present chapter.

Section B is an attempt to project some of INTELSAT's more likely futures. A particular effort will be made to define statutes, roles, activities, policies and structures for INTELSAT that will assure it a viable long-term existence in facing the challenges that have already been outlined. This will not necessarily be the easiest or the most comfortable evolution for the organization to take in the near term. Short-run sacrifices will be necessary to help INTELSAT realize its preferred future into the twenty-first century. Section C contains a summary.

A. *Conceptual Frameworks Within Which INTELSAT Can be Analyzed and Evaluated*

1. *Functionalist theory*

As was noted in Chapter III, functionalist theory, stated simply, argues that cooperation among nations in essentially technical, non-controversial, non-political undertakings can expand gradually so that it displaces conflict-building political activities in the international arena. A particular focus of this point of view is not merely the nature of the cooperative activities that take place but the institutional mechanisms that evolve to accommodate them.

INTELSAT would certainly appear to have fulfilled a functionalist role among nations, particularly during its initial years. Pelton (1974) reviews INTELSAT's early experience favorably within the functionalist perspective, noting that it attained its greatest success in relatively non-controversial areas considered important by all of its members, such as establishing the initial technical and operational parameters of the global system. Snow (1976, pp. 133–45), without explicit use of functionalist theory and employing a more economic frame of reference, concluded in 1976 that much of INTELSAT's first decade of successful operation could be attributed to its avoidance (whether by chance or by design) of divisive political and distributive issues, a task at which many other international organizations have failed. Pelton (1974, p. 158) mentions EURATOM and INTERSPUTNIK in this latter context.

It is misleading, however, to suggest that INTELSAT's early years were free of political undercurrents or explicit policy choices. The two most politically divisive issues of its first decade were, in retrospect, the renegotiation of the agreements during 1969–71 and the internationalization of the bidding procedures for the INTELSAT IV series during roughly the same time period. Chapter IV of this study catalogues and examines the major challenges to INTELSAT today, most of which are likewise overtly political. Let us review them to determine the extent of their political content.

The separate-systems issue, while ostensibly an economic one relating to the financial health of INTELSAT after competitive entry, has been politicized from the very first, and little or no hard economic analysis has surfaced. Fiber-optic cables, by contrast, since they intersect minimally with INTELSAT's judicial competence and decision-making apparatus, are viewed in a more objectively economic framework. INTELSAT does not dispose of the legal or political means to oppose the new generation of fiber-optic cable systems even if it wished to do so.

Comsat's role in INTELSAT has also been a political one at times. The sheer magnitude of its early role in INTELSAT – as a veto-wielding signatory, as the system manager and as the supplier of personnel for INTELSAT's secretariat – was in and of itself a political issue, regardless of how precisely the corporation's technical, legal, financial or managerial role was defined. Comsat's challenges today, such as pricing flexibility, direct access to the space segment by U.S. competitors and competition from cables, are microcosmic metaphors of INTELSAT's own political difficulties.

Deregulatory trends in several of INTELSAT's industrialized signatories – particularly in the United States and Japan – have for the most part been pragmatic and economically based. In others, such as the United Kingdom, change has come for manifestly political reasons. The PTTs of certain countries, such as France and Australia, have persuasive political arguments for opposing the liberalization of their telecommunications sectors, while Canada and West Germany find any change difficult because of the delicate political and constitutional balance of federal-state relations. While deregulation has had minor economic effects on INTELSAT – such as pressure to bring costs closer to prices – these impacts have doubtless been outweighed by the political fallout of deregulation. Similarly, the effects of increasing competitive pressure on INTELSAT have raised the specter of less INTELSAT aid to its developing members. This, in turn, may well exacerbate political tensions between industrialized and developing signatories if and when individual countries, by acting in their own narrow interests, choose to break the rules of the INTELSAT »cartel«[1] and thereby threaten its stability and existence.

Certain of INTELSAT's present challenges can still be insightfully interpreted within the functionalist framework. The shortage of launch vehicles, for example, is an inherently technological problem that might well expand the ambit of international cooperation into non-political operational and technical areas. Certainly such an expansion would come to pass if INTELSAT were to use a Soviet Proton or a Chinese Long March launcher. Likewise, INTELSAT's experience among the community of international organizations has tended to confirm the functionalist hypothesis. Each memorandum of understanding with another organization, for example, has increased the area of ostensibly non-political international cooperation in which INTELSAT engages. Unfortunately most of this cooperation has occurred among Western industrialized countries rather than between East and West or North and South.[2]

We must conclude that the functionalist framework is a tempting but incomplete one within which to evaluate INTELSAT. Although the organization's field of activity in technical, non-political enterprises has increased greatly in recent years, this expansion has not reduced the areas of political friction. Indeed, INTELSAT's political conflicts have increased *pari passu*.

1 See footnote 18 of Chapter IV for an interpretation of INTELSAT as a cartel.

2 INTELSAT's aid programs for developing countries, outlined in Section G of Chapter IV, constitute the major North-South component of such cooperation.

2. *Cooperation and social economy*

Voge (1986) has dealt thoughtfully with cooperation, arguing that it is preferable in many instances to competition. Among the benefits of cooperation are smaller risks of a proliferation of technical standards and protocols; less chance of »uneconomic entry« into an unsustainable natural monopoly; and, somewhat paradoxically, less chance of anticompetitive collusion by large firms to foreclose the very benefits that the competitive marketplace is supposed to provide. In addition, cooperation is seen as a more effective approach to decentralizing telecommunications and the media, a priority in France today.

The nineteenth-century English cooperatives, as well as German economic thinkers such as Adolf Wagner of the same general period, are, as we have previously noted, the intellectual antecedents of the strain of social economic doctrine known as *Gemeinwirtschaftslehre* in West Germany. We now consider this school of thought as a second framework for evaluating INTELSAT. It provides a socialist but non-Marxian rationale for furnishing vital public services such as telecommunications under a framework of government monopoly.

Thiemeyer, the principal contemporary exponent of this view, conceives of the social economic enterprises as a »stopgap« or substitute for private ownership and competition when a political consensus emerges to the effect that the market fails to perform satisfactorily in a given situation (*see* Thiemeyer 1970). It is noteworthy that the question is not whether there is an objective malfunction of market forces but whether such a malfunction is perceived to exist by a majority of citizens. Thiemeyer also contends that any definition of the »public interest« or »common good« *(Gemeinwohl)* should be made on a case-by-case basis, once again taking political perceptions into account.

In a recent article dealing directly with telecommunications, Thiemeyer (1983) forcefully rebuts arguments for liberalization put forth by advocates of the neoclassical point of view (*see* Section A(5) below). He not only rejects marginal-cost pricing but also opposes any linking of prices to costs by regulatory rules, primarily because of the difficulty in econometrically determining costs and the large bureaucracies needed to formulate and enforce pricing rules.[3] In addition, he justifies the patterns of cross-subsidization

3 Various theoretical and practical objections to marginal-cost pricing are also put forward by Copeland and Severn (1985). It is noteworthy that Knieps, Müller and von Weizsäcker (1981), analyzing the Deutsche Bundespost from a neoclassical perspective, argue against Ramsey pricing with much the same reasoning that Thiemeyer (1983) uses against marginal-cost pricing: econometric difficulties in estimating the relevant price elasticities of demand, and political complications.

implicit in any politically determined pricing structure as a way of achieving various social objectives, such as »regional equalization of burdens.«
The classic criticism of *Gemeinwirtschaft* is that of von Mises (1981, translation of 1932 German edition).
It should be noted that although the Bundespost is occasionally referred to as a »social economic enterprise« (*gemeinwirtschaftliches Unternehmen*), the core representatives of this concept are corporations owned and controlled by trade unions rather than the federal government. The recent bankruptcy of Neue Heimat, a union-owned enterprise that was the largest housing construction firm in Europe, dealt a serious blow to the appeal of such concerns.
There is an affinity between arguments such as these to provide a public-monopoly »stopgap« for vital services and the conceptually much more secure notion of prohibiting competitive entry in order artifically to »sustain« an unsustainable natural monopoly. Snow (1984) has commented on this similarity of public-policy prescriptions in these two very different approaches to public regulation. There is no question of one approach influencing the other, since they are based on methodologies and philosophical outlooks that are mutually quite incompatible. For example, the sustainability of monopoly literature, as we have observed in Section C(6)(b) of Chapter I, is strongly grounded in neoclassical economic theory, while even modern adherents of *Gemeinwirtschaftslehre*, such as Rittig (1977), have complained of its lack of rigor and its incomplete basis in theory and fact. What is at work here is the same underlying technological dynamic we have commented on throughout this study. In the case we are now examining, that process appears to have given rise to a similar policy prescription in two widely diverse approaches to political economy.
INTELSAT's main benefit from cooperative principles has been in its financial operation. As a financial cooperative of owners and users, it has removed the harmful structural dichotomy between these two groups that has been so divisive in most domestic telecommunications markets. In addition, Article 7 of the Operating Agreement lays down a procedure whereby signatories desiring an excess of investment quota over usage (perhaps to earn profits they are denied as domestic PTTs) can purchase additional shares from other signatories (perhaps those not yet in a position to contribute significantly to financing the INTELSAT global system).[4]
Has INTELSAT been a »stopgap« solution, filling a niche that has been left empty by the competitive market? This was evidently the case during its early

4 Johnson (1987, pp. 23–24) notes that Comsat is an enthusiastic »oversubscriber« to INTELSAT ownership because of incentives relating to its rate-base regulation by the FCC.

years of operation. Until well into the 1980s it was economically and institutionally unthinkable to consider private, competitive, market-based alternatives to creating a global, commercial satellite system. Certainly, in the phraseology of social economic thought, there was a political consensus that the market was unable to perform that function. This consensus could not have been expressed any more forcefully than by the formation of an international organization now consisting of 113 countries and over $1,600 million in net investment. Yet that consensus is now unraveling; adherents of social economic thought, who maintain that interventions in the market economy must be considered on a case-by-case basis, should accordingly reassess their conclusions. The same prescription applies, *mutatis mutandis*, to those who hold that cooperation is superior to competition in the telecommunications sector.

3. *Political economy of deregulation*

This approach, which incorporates normative concerns into regulatory analysis by taking the political environment explicitly into account, has been extensively applied to telecommunications. Olson's (1982) book is the seminal reference for this school of thought. He posits and tests several hypotheses about the role of interest groups, and their structural conflict with the overall public interest, in industrialized Western countries. Researchers in this area have asked and attempted to answer the following basic questions in many different ways. How are small, well-organized interest groups in industrialized countries able to frustrate the implementation of regulatory changes that would demonstrably increase aggregate social welfare? By what political actions and mechanisms do such groups acquire that power? What arrangements can be made to bring about a transition to a condition of maximum social welfare by paying off prospective losers with part of the benefits to be realized by prospective gainers?

Noll (1983, 1986b) has provided a theoretical basis for examining how different political institutions affect regulatory policy and has laid the groundwork for international comparative studies in this area. He argues that an institutionalist context, which takes the political environment explicitly into account, is preferable to democratic, statist or pluralist paradigms in explaining the international variation of regulatory structures, policies and outcomes. Empirically, he has examined political phenomena, such as the way in which representatives are elected to the national parliament and the degree of independence of the parliament with respect to the executive branch, as vehicles for explaining cross-country variation in the nature, pace and extent of telecommunications liberalization. He has concluded as follows:

> [T]he problem of developing effective national interventions by governments at the level of industries and companies is more severe in relatively self-sufficient, diverse societies and in nations with political structures that tend to fragment decision making among political leaders with separate constituencies. Effective regulation or industrial policy, therefore, ought to be more difficult in Britain or the United States than, say, in Japan, with West Germany and France in intermediate positions (Noll 1986b, p. 62).

Other studies in the political economy of deregulation have illuminated the regulatory process in the United States. Noll and Owen (1983), who analyzed interest-group politics in several U.S. industries, argue in favor of competition and against regulation on general principles. Derthick and Quirk (1985) conclude that the »politics of ideas« has been the primary force in bringing about deregulation in the U.S. civil aviation, trucking and telecommunications industries. In addition, they maintain that the patient and politically astute advocacy of deregulation by government, business and academic economists has played a vital role in its adoption.

How, if at all, is the political economy of deregulation associated with INTELSAT's present challenges? At the outset, it is clear that the frame of reference and analysis of this approach must be altered to suit the unique features of INTELSAT. Until now, that framework has been applied to the domestic political arena rather than to international organizations. In addition, the interest groups, and their interests, must be identified. To a certain extent, indeed, INTELSAT's cooperative structure has eliminated or at least masked the interest-group conflicts that have been so counterproductive in the domestic regulatory politics of many countries. A little thought, however, can suggest the outlines of an approach to using the political economy of deregulation as a way of evaluating INTELSAT's present policy choices.

Let us assume for the moment that if INTELSAT adopts certain changes it will become a stronger and more economically viable organization in the long run. We shall discuss some of these changes in Section B below. For the sake of argument let us assume that three such changes are (1) the admission of competition from most or all separate international satellite systems having the financial, technical and managerial resources to enter operation; (2) the amendment of INTELSAT's agreements to give it greater pricing flexibility to meet such competition; and (3) the curtailment or elimination of development aid programs due to the disappearance of implicit subsidies in the new tariff structure. What interest-group conflicts would arise as a result of this transition to what we have assumed is a state of greater overall social welfare?

The analysis requires identification of the prospective losers and gainers. The gainers would clearly include private firms establishing separate satellite systems, as well as launch and earth-station entrepreneurs. End users in all countries would also benefit from the diversity of new services that competition inevitably brings, as well as from the consequent downward pressure on prices. It is also arguable that private or recently privatized INTELSAT signatories, such as Comsat, BT, NTT and Telespazio, would gain on balance as commercial organizations retailing competitively priced INTELSAT circuits to the public in competition with fiber-optic cables. Another major constituency benefiting from the change would be the various executive and regulatory agencies of the U.S. government that have lobbied so long for such a change. Who would lose? Developing countries might be the most powerful constituency affected in this manner. Cost-based pricing would increase prices to most of the Third World. In addition, it would eliminate the funds that have in the past been available for the many INTELSAT aid programs benefiting LDCs. INTELSAT's full-time staff might also perceive its interests as being adversely affected by the evolution to a more »corporate« structure, especially to the extent that Averch-Johnson behavior (*see* Averch and Johnson 1962) might have overcapitalized the organization in the past. In addition, traditional PTT signatories not functioning under competitive or commercial conditions might find various disadvantages in dealing with a new INTELSAT subjected to competitive and pricing discipline.[5] Finally, operators of fiber-optic cable systems might well see their interests threatened if INTELSAT were to price its circuits closer to their (presumably low) marginal cost.

What mechanisms exist for prospective gainers to compensate prospective losers and thereby to facilitate the transition? First, the interests of developing countries could be served by establishing a telecommunications development fund supported by the members of INTELSAT and by its satellite competitors. This fund could also be contributed to by outside agencies, such as the ITU and the World Bank. Cable operators, to the extent their interests were actually threatened, might be guaranteed minimum circuit-loading shares. It seems more difficult to conceive of ways in which the real or imagined psychic and institutional losses inflicted on INTELSAT's staff or its non-commercial signatory PTTs might be redressed.

5 This possibility is suggested by the historical disinclination of traditional PTTs to enter into substantive negotiations with private U.S. carriers for international telecommunications facilities.

This example illustrates the very political process of negotiation and compromise required to bring about a hypothetical change that, although it is on balance superior to the status quo, might easily be frustrated by a coalition of powerful groups that see their own interests threatened.[6] The heart of the matter is the redistribution of the costs and benefits occasioned by change. Although the theoretical precepts and methods of the political economy of deregulation school have not yet been applied to the case of INTELSAT, there would appear to be ample opportunity for scholars to do so.

Until now, change in INTELSAT has not come painlessly. There has not been an easy accommodation of burdens and benefits among its various constituent groups. As the separate-systems issue has shown, opposing sides have hidden behind non-negotiable positions and have avoided discussing in particular the substantive economic issues at the core of their dispute. Perhaps the insights we have referred to in this subsection can help to make the issues in such changes more transparent and the parties involved more disposed to negotiate and compromise.

4. *Collective-goods model*

Martinez[7] outlines the collective-goods perspective of microeconomics and political science as it applies to satellite communications. He emphasizes the dichotomy between the radiofrequency spectrum (which is »indivisible« and »nonexcludable« and thereby a collective good) and the more recently exploited geostationary orbit (which, being both divisible and excludable,[8] is therefore not a collective good). The ITU and other international organizations and treaty arrangements were created as a natural outcome of the collective-goods aspect of the spectrum before the advent of satellites. The addition of the geostationary orbit to the storehouse of communications resources is held to cause significant political tensions in the existing international regulatory regime. Since the goal of any state is »its security and survival in an anarchic and threatening international system,« there is »an inherent tension between a state's perceived need to cooperate and its

6 Noam (1986b) has used the term »postal-industrial complex« to characterize the constellation of forces in Western Europe opposed to telecommunications liberalization.

7 See Martinez (1985, pp. 149–61) for a concise application of the collective-goods (public goods) literature to satellite communications. His footnote 1 (pp. 159–60) is a useful reference to several basic sources.

8 Excludability in this sense is taken to mean the expense and difficulty of reaching the geostationary orbit relative to accessing the radiofrequency spectrum.

instinctual guarding of sovereign prerogatives.« As a result, a state will participate in international organizations such as the ITU only »if the perceived benefits outweigh imposed dependency costs« (Martinez 1985, p. 150).

This careful calculus has been upset by the asymmetry noted above between the properties of the spectrum and those of the geostationary orbit. Thus, existing international machinery in the ITU, the U.N. Outer Space Treaty and elsewhere is not fully equipped to address and mediate conflicting claims to the geostationary orbit.

In the case of a collective (indivisible, non-excludable) good, »rational, self-interested individuals will not act to achieve their common or group interests. The difficulty in excluding non-contributors from enjoying the benefits of a collective good leads to that good's underproduction due to the so-called ›free-rider‹ problem« (Martinez 1985, p. 153; emphasis in original omitted).

This in turn gives rise to the organizational forms and sanctions provided by INTELSAT and the ITU. New technology, nonetheless, has bestowed separate international satellite systems with qualities primarily of private rather than collective goods:

> Technology is changing the incentives for forming common-user networks in favor of dedicated national or regional systems. These represent an excludable good of increasing power value for the few countries with the means for direct access to the orbit resource .... In essence, technology has created a free-rider alternative. Excludable regional, national, and dedicated ... satellite communications networks allow the satellite operator to receive the benefits without the dependency costs incurred as a participant in the global network. INTELSAT is becoming the satellite network for countries that cannot afford direct access, as wealthier countries with the technological and financial means establish their own national and regional systems (Martinez 1985, p. 157).

Two observations are appropriate at this point. First, Martinez's treatment seems more directed to the domestic and regional systems already coordinated with INTELSAT under Article XIV procedures than to the privately owned, largely U.S. systems presently clamoring for such coordination. Second, the market-like mechanisms for allocating spectrum and orbit discussed in Section C(3) of Chapter III seem particularly germane as ways of overcoming the excludability property of the orbit that faces developing countries.

The collective-goods model is a useful guide for evaluating possible reforms in the existing INTELSAT regime as a way of forestalling »a disintegration of INTELSAT into a plethora ... of smaller, less efficient satellite systems« (Martinez 1985, p. 159). One suggestion is that in order to reduce dependency costs of sovereign states in supranational organizations, contributions to a collective good can be encouraged by making »the good appear as private as possible« (Martinez 1985, p. 158). This can be done, for example, by breaking up the ITU into regional organizations and managing the orbit and spectrum resources on a regional basis. In addition, pricing flexibility for INTELSAT will surely bring about a certain regionalization of that organization as circuits become tariffed on a route-by-route basis according to demand and other market criteria.

5. *Neoclassical economics*

Most of the economic analysis used in this study has been explicitly or implicitly neoclassical in nature. This approach emphasizes the optimization (maximization or minimization) of certain key economic variables (such as utility, profit, cost or revenue) subject to constraints on the availability and behavior of various other economic quantities (as expressed by the utility function, the cost function, the profit function, the demand function and so on). In addition, it is fair to say the neoclassical approach more often than not favors competition and market processes as superior methods of carrying out the optimization referred to, although regulatory economics and industrial organization in the neoclassical tradition extensively investigate cases of »market failure« in which competitive forces do not optimize properly.

The most novel and insightful application of neoclassical theory discussed in this study has been that of natural monopoly, sustainability of natural monopoly and contestability of market structure introduced in Section C(6) of Chapter I. In particular, cost minimization for producing a given combination of services has been a criterion for establishing the property of natural monopoly. Contestability has been described as a method of approximating the optimal welfare properties otherwise possible under the perfect competition model. Throughout this study we have shown how INTELSAT fits into these explanatory constructs, and we have interpreted its past behavior and preferred policy future accordingly.

There is, to be sure, no universal acceptance of the neoclassical approach to economics and to political economy. One criticism, in fact, is to the effect that neoclassical theory is entirely descriptive and analytical (positive) rather than prescriptive (normative).[9] This objection is perhaps the most telling one

9 Zajac's (1978) monograph is a useful and non-technical introduction to the equity implications of telecommunications policy. Wenders (1987, pp. 63–92) sets forth the normative implications of telecommunications pricing in a recent book-length treatment of telecommunications economics.

leveled by adherents of the social economy school, for example. Even from within the neoclassical framework there are various criticisms of particular policy conclusions. For example, Park and Mitchell (1986) have concluded, on the basis of extensive experience in telecommunications pricing, that usage-sensitive pricing is in general less efficient than flat-rate pricing in telephone networks, even though a strong theoretical case has traditionally been made for the former. Likewise, Copeland and Severn (1985) challenge the widespread neoclassical conviction that structural deregulation of telecommunications is required to maximize social welfare. Applying their argument to intra-LATA markets in the United States, they maintain that economic efficiency need not be maintained at the cost of »significant departures from historic pricing principles« and »wholesale competitive entry into intra-LATA markets.« In contrast to the great majority of pricing theorists, they present algebraic and diagrammatic arguments against the use of Ramsey and marginal-cost pricing and in favor of average-cost or unbundled prices.

The shortcomings of neoclassical economics as it applies to telecommunications can best be addressed within rather than outside of that tradition. A recent research workshop investigating economic policies relating to the deregulation of telecommunications in seven industrialized countries found that several aspects of neoclassical analysis are in need of internal reform (*see* Snow 1986c, pp. 280–281). First, a number of technical objections to neoclassical analysis emerged. For example, there is considerable hesitation about the dependability of the underlying cost and production functions that can be derived from inaccurate and incomplete data. Such data are often the only kind available from communications carriers, who usually have strategic incentives to distort or conceal costs. Even with the best of data, modern cost-accounting methods have yet to come to terms with many practical difficulties in the measurement and allocation of telecommunications costs. In addition, we have noted that externalities – economic relationships not captured by the price system – can cast doubt on the conclusions of neoclassical welfare analysis. Finally, as Lamberton (1986) has noted in the case of Australia, there are social and perhaps psychological considerations mitigating the force of certain conclusions derived from neoclassical analysis. Noll (1986b) touches on many of the political qualifications to the neoclassical approach. These constitute the core of the political economy of deregulation already mentioned. Tightly organized interest groups can exploit peculiarities of the political process to prevent the implementation of economic policies that are demonstrably welfare-superior to the status quo from an aggregate point of view. Even in using the term »aggregate,« it should be noted, we beg the important question of how to define overall social welfare (the »common good«) or whether it even exists in objective reality.

How can INTELSAT's experience and prospects be evaluated using neoclassical analysis? Examining its first decade of operation, Snow (1976, pp. 133–49) concluded that INTELSAT had performed well when judged on the criterion of maximizing the welfare of consumers. After more than two decades of operation, however, INTELSAT now appears to be at a particularly critical watershed. Technological forces have redefined and eroded the natural-monopoly core within which INTELSAT can provide certain combinations of services more cheaply alone than two or more satellite systems can together.

In reviewing the five frameworks of analysis and evaluation presented above, we can conclude that each has a particular relevance to INTELSAT. The most broadly appropriate and applicable model, however, is clearly that of neoclassical economics. Indeed, the contributions of the other four – functionalist theory, cooperation and social economy, the political economy of deregulation and the collective-goods model – can usefully be interpreted as refinements, qualifications or extensions within the overall economic superstructure. We shall now turn to prospects for INTELSAT's future, making use primarily of the economic criteria of evaluation developed so far in this study.

B. *Prospects and Scenarios for INTELSAT's Future*

A number of scholars and policymakers have speculated about the future course of INTELSAT. Their observations are instructive as we seek to identify a preferred future for the organization.

Pelton and Alper,[10] looking a century into the future, concluded as follows:

> By the year 2085, INTELSAT can certainly be expected to have evolved and changed in many ways. It could become a more decentralized system, connecting different types of regional consortia and perhaps providing dedicated subsystem capabilities to corporations, national governmental agencies, and regional and international organizations. The key to INTELSAT's survival and growth over the next hundred years is surprisingly much the same as it has been for the last 20 years: Namely, to develop new, more cost-effective technologies; to develop new

10 At the time, Pelton was director of Strategic Policy Formulation at INTELSAT and Alper was president of Comsat's World Systems Division.

and innovative services while reducing the cost of conventional services; to remain as an effective global common denominator, responsive to the needs and demands of all countries and all peoples; and to constantly evolve new strategies and maintain institutional flexibility to respond to a rapidly changing economic, social, and political global environment (Pelton and Alper 1984, p. 378).

Charyk and Goldstein,[11] considering the more immediate needs of INTELSAT, expressed the following observations:

Changes initiated within INTELSAT several years ago have begun to bear fruit as the number of service alternatives for users explodes, with significantly increased technical and price options. In fact, the vitality and viability of the INTELSAT system, and its value to the variety of users it serves in developed and developing countries, rest precisely on its ability to offer shared resources for a variety of applications. It is to be hoped that INTELSAT's members, while propelling the organization into the emerging global network, will similarly focus on the wider variety of services that the integrated international cooperative can make possible. ... While establishment of more networks may look attractive, unneeded duplication of facilities to provide limited services to a few users can only be more expensive, especially to the majority of users who can least afford it. It is new technologies and services, not networks, that are needed (Charyk and Goldstein 1984, p. 385).

Finally, Bruce, Cunard and Director, after a thorough examination of domestic and international telecommunications policy in the United States, arrived at the following conclusions about INTELSAT's future role:

INTELSAT ... might become more active as a coordinator of numerous regional, nationally based systems; it might reduce its role as a procurement agent for international satellite services directly from manufacturers. INTELSAT also could take an active role in devising and overseeing a multinational mechanism (explicitly not using embedded rate subsidies) to subsidize satellite services to less-developed regions and less-used routes around the world (Bruce, Cunard and Director 1986, p. 280).

Using these insights and the observations of this study, let us now assemble the most important elements of a preferred future for INTELSAT – one that will keep the organization economically, technologically and politically viable well into the twenty-first century.

11 At the time, Charyk was chairman and chief executive officer of Comsat and Goldstein was president of Comsat.

1. *Competition, pricing flexibility and a core of natural-monopoly services*

INTELSAT's single most important challenge in the face of inevitable competitive entry by separate satellite systems will be to adapt to that competition successfully and, indeed, to use competition to its own advantage. According to the arguments of Hayek (1968) and the substainability-of-monopoly literature, competition will disclose or »discover« the core of natural-monopoly services that INTELSAT can produce by itself more cheaply than can two or more systems jointly. This is the domain in which the establishment of separate facilities is »uneconomic.« In the case of unsustainable natural monopoly, incentives for uneconomic entry exist. INTELSAT's best defense against uneconomic entry – one that is in the interest of all end users as well – is to price flexibly according to marginal cost on a service-by-service basis, supplementing revenues by using demand considerations (such as Ramsey pricing) if necessary. INTELSAT may have to amend its agreements to achieve this flexibility. In addition, regulations and accounting safeguards may well be necessary to ensure that pricing flexibility does not offer INTELSAT the opportunity to violate the antitrust laws of any of its members by practices such as predatory pricing.

New technologies may have shrunk the extent of INTELSAT's core of natural-monopoly services during recent years, but there is no particular indication that this is an irreversible trend. The advent of integrated switched digital network (ISDN) technology and services, for example, may well expand the natural-monopoly output domain for satellite services and telecommunications generally.[12] In any event, competition is by far the most efficient method of determining where INTELSAT's cost and technological advantage lies.

12 Quite diverse views exist as to the motives behind ISDN introduction and the extent of its eventual integrative effect on global telecommunications. Some hold that the alleged scale and scope economies of ISDN will usher in a new era of network harmony and unity, virtually rolling back decades of deregulation. *See* Neumann and Schnöring (1985), who dismiss fears about the »social governability« and »micro- and macroeconomic rationality« of ISDN. Noam, by contrast, has attacked ISDN in the following terms:

> The notion that in the age of information all communications flows ... will pass through one (or at most a few) »streamlined« communications links of a single organization seems hard to entertain on technical, economic or political grounds, except by reference to institutional conservatism and the present balance of political powers. But these conditions may not prevail in the long run; like a Greek tragedy, the unified system of hegemonial ISDN will unravel, because it reflects the needs of a bygone era (Noam 1986a, pp. 21–22).

See Tudge (1987) for a statement by an INTELSAT official regarding ISDN.

2. *INTELSAT as a source of developmental assistance*

Commercial pressures of the kinds just described may eliminate any surplus INTELSAT might have enjoyed in the past to assist developing countries in realizing the benefits of satellite communications. INTELSAT's experience and expertise in that area, however, would certainly appear to place it advantageously in seeking outside funds to continue such aid. Obvious sources include the World Bank, the ITU and various national development funding agencies such as the U.S. Agency for International Development (AID). INTELSAT could perform these services on a cost-plus basis in much the same way it earns a fourteen percent return by leasing its satellite circuits. In addition, there would be numerous synergies and complementarities involved in providing this aid. Continuing assistance by INTELSAT would encourage LDCs to use and to expand their use of satellite communications and would assure INTELSAT of ongoing support in the developing world.

3. *Advanced technology research*

INTELSAT's research and development to date have been quite strictly focused on the organization's own needs. The INTELSAT of the future might well undertake more advanced, even exotic research for external entities on a cost-plus basis, building on its two decades of research and operational experience in satellite communications. Topics of inquiry in the immediate future could include space platforms, antenna farms, teleports, intersatellite links and fiber-optic techniques.[13] Research into implications of satellite communications for the social sciences and law should also be considered. Areas of investigation extending into the twenty-first century are limited only by the imagination.

4. *INTELSAT as a global coordinator of all satellite systems*

INTELSAT is excellently positioned to monitor not only its own satellites but also those of all regional and international systems. It could serve as a kind of global switchboard and in so doing could gain and provide economic and technical insights into the optimal coverage of the earth's surface by various

13 See Pelton and Alper (1984, pp. 369–70) for the conceptualization of an INTELSAT antenna farm in the year 2010.

kinds of systems. This empirical research could supplement the information provided by the competitive process in discovering and monitoring any changes in INTELSAT's natural-monopoly core.

5. *INTELSAT as a diplomatic facilitator and mediator*

As an international organization in its third decade of operating a commercial enterprise, INTELSAT is strategically located between the sovereign states that establish and regulate the conditions under which international communications is conducted and the private industries and firms that develop and supply the technology and systems making communications possible. As the institutional landscape of international telecommunications becomes more crowded and complex, INTELSAT's collective wisdom and experience might well be of value in keeping that landscape orderly and uncluttered. INTELSAT might, for example, assist other countries or groups of countries in drafting new treaties or agreements. It might provide tribunals to mediate disputes or to enforce decisions already reached.

6. *New entrepreneurial areas*

As we noted in Section B(1) above, future technological developments might be such that INTELSAT's natural core of profitable activities expands rather than contracts. If this is the case, INTELSAT could enlarge its mandate by appropriately amending its agreements. For example, it may wish to merge its activities horizontally into specialized services. One potential area is aeronautical satellite services. These have been technologically feasible for twenty years, but for a number of financial and institutional reasons, they have not yet been offered commercially. INTELSAT could also acquire INMARSAT, for which it provides a considerable number of circuits. INTELSAT might even wish to acquire certain undersea cable systems, if economies of scale and scope made that a logical initiative, or to integrate vertically »upstream« by creating its own launcher subsidiary. All of these areas of potential expansion, it will be noted, are inherently transnational. For political reasons there would appear to be little opportunity for INTELSAT to integrate »downstream« into the domestic telecommunications sectors of individual countries, although acquiring the various ground segments accessing INTELSAT, for example, would certainly contribute to a better global optimization of the system.

C. *Summary*

Rumors of INTELSAT's ill health or impending demise are exaggrerated. Fiber-optic cables and competing international satellite systems will not replace INTELSAT, any more than television has replaced motion pictures or video tape recorders will replace television. In global telecommunications, exponential increases in capacity and technological sophistication have been accompanied by exponential decreases in costs. These quantitative indices, however, have brought with them a qualitative breadth and variety of telecommunications services that have fueled a transformation of the economies of the industrialized world. If, as has been contended, the major acitivity of the global economy has shifted from manufacturing to information, then telecommunications is the catalyst of that transformation, just as the steam engine and other techniques aided the shift from agriculture to manufacturing in the Industrial Revolution two centuries ago. Section B above suggests a number of adaptations that INTELSAT might make to the altered circumstances it now faces and will continue to confront as a result of these underlying technological transformations. Some changes, such as pricing flexibility and toleration of competition, are essential. Others might be advisable as future conditions dictate or permit.

In short, INTELSAT has the institutional flexibility to survive, to respond to and even to thrive in the technological and economic environment of the twenty-first century. Its successful response to the challenges of the past is the strongest indication of its ability to realize the even greater opportunities of the future.

Bibliography

Alper, J.R. and J.N. Pelton (eds.)
1984 *The INTELSAT Global Satellite System* (Progress in Astronautics and Aeronautics, vol. 93). New York: American Institute of Aeronautics and Astronautics.

Averch, H. and L.L. Johnson
1962 Behavior of the Firm Under Regulatory Constraint. *Am. Econ. Rev.* 52(5):1052–69.

Bailey, E.E. and W.J. Baumol
1984 Deregulation and the Theory of Contestable Markets, *Yale J. Reg.* 1(2):111–38.

Bailey, E.E. and A. Friedlaender
1982 Market Structure and Multiproduct Industries. *J. Econ. Lit.* 20(3):1024–48.

Bargellini, P.L. (ed.)
1974 *Communications Satellite Systems*. Cambridge (MA): MIT Press.

Baughcum, A.
1986 Deregulation, Divestiture, and Competition in U.S. Telecommunications: Lessons for Other Countries, in *Marketplace for Telecommunications: Regulation and Deregulation in Industrialized Democracies*, edited by M.S. Snow, 69–105. New York: Longman.

Baumol, W.J.
1982 Contestable Markets: An Uprising in the Theory of Industry Structure. *Am. Econ. Rev.* 72(1):1–15.

Baumol, W.J. and D.F. Bradford
1970 Optimal Departures from Marginal Cost Pricing. *Am. Econ. Rev.* 60(3):265–83.

Baumol, W.J. and Y.M. Braunstein
1977 Empirical Study of Scale Economies and Production Complementarity: The Case of Journal Production. *J. Pol. Econ.* 85(5):1037–48.

Baumol. W.J., J.C. Panzar and R.D. Willig
1982 *Contestable Markets and the Theory of Industry Structure*. New York: Harcourt Brace Jovanovich.
1986 On the Theory of Perfectly-Contestable Markets, in *New Developments in the Analysis of Market Structure*, edited by J.E. Stiglitz and G.F. Mathewson, 339–65. Cambridge (MA): MIT Press.

Boiteux, M.
1971 On the Management of Public Monopolies Subject to Budgetary Constraints. *J. Econ. Theory* 3(3):219–40.

Brock, W.A.
1983 Contestable Markets and the Theory of Industry Structure: A Review Article. *J. Pol. Econ.* 91(6):1055–66.

Bruce, R.R., J.P. Cunard and M.D. Director
1986 *From Telecommunications to Electronic Services: A Global Spectrum of Definitions, Boundary Lines and Structures*. Boston: Butterworth Legal Publishers.

Business-Higher Education Forum
1986 *Space: America's New Competitive Frontier*. Washington, D.C.

Caplan, L.A.
1986 The Case For and Against Private International Communications Satellite Systems. *Jurimetrics J.* 26(2):180–201.

Charnes, A., W.W. Cooper and T. Sueyoshi
1985 A Goal Programming/Constrained Regression Review of the Bell System Breakup (paper presented at the meetings of the European Operations Research Society). Bologna (Italy).

Charyk, J.V. and I. Goldstein
1984 INTELSAT and its Role in the Emerging World Telecommunications Systems, in *The INTELSAT Global Satellite System*, edited by J.R. Alper and J.N. Pelton, 379–85. New York: American Institute of Aeronautics and Astronautics.

Clarke, A.C.
1945 V–2 for Ionosphere Research and Extraterrestrial Relays. *Wireless World*. Feb.:58.

Codding, G.A., Jr. and A. Rutkowski
1982 *The International Telecommunication Union in a Changing World*. Dedham (MA): Artech House.

Colino, R.R.
1983 Statement of Richard R. Colino, Director General-Designate of INTELSAT, before the Subcommittee on Arms Control, Oceans, International Operations and Environment, U.S. Senate Foreign Relations Committee (mimeo). Washington, D.C.
1984 The INTELSAT System: An Overview, in *The INTELSAT Global Satellite System*, edited by J.R. Alper and J.N. Pelton, 55–94. New York: American Institute of Aeronautics and Astronautics.

Copeland, B.L., Jr. and A. Severn
1985 Price Theory and Telecommunications: A Dissenting View. *Yale J. Reg.* 3(1):53–85.

Corbo, V. and M. Munasinghe
1978 The Demand for International Telecommunications Services and Their Policy Implications: The Case of Canada. *Telecommunication J.* 45(9):493–503.

Coursey, D., I.R. Mark and V.L. Smith
1984 Natural Monopoly and Contested Markets: Some Experimental Results. *J.L. & Econ.* 27(1):91–114.

Courteix, S.
1984 EUTELSAT: Europe's Satellite Telecommunications, in *Regulation of Transnational Communications* [1984 Michigan Yearbook of International Legal Studies], edited by L.J. Anderson, 85–101. New York: Clark Boardman.

Cowhey, P.F. and J.D. Aronson
1985 The Great Satellite Shootout. *Reg.* 9(3):27–35.

Crutchfield, J.
1961 An Economic Evaluation of Alternative Methods of Fishery Regulation. *J.L. & Econ.* 4(2):139–43.

Derthick, M. and P.J. Quirk
1985 *The Politics of Deregulation*. Washington, D.C.: Brookings.

Elliot, D.W.
1987 PPAC Private Cable Between the U.S. and Japan, in *Proceedings: Pacific Telecommunications Conference 1987*, edited by D.J. Wedemeyer and M.S. Bissell, 61–63. Honolulu (Hawaii): Pacific Telecommunications Council.

Evans, D.S. and J.J. Heckman
1983a Multiproduct Cost Function Estimates and Natural Monopoly Tests for the Bell System, in *Breaking up Bell: Essays on Industrial Organization and Regulation*, edited by D.S. Evans, 253–82. New York: North Holland.
1983b Natural Monopoly, in *Breaking up Bell: Essays on Industrial Organization and Regulation*, edited by D.S. Evans, 127–56. New York: North Holland.
1984 A Test for Subadditivity of the Cost Function with an Application to the Bell System. *Am. Econ. Rev.* 74(4): 615–23.

Fishman, W.L.
1984 Competing with INTELSAT. *Telecommunications* 18(6): 112–16.

Galloway, J.F.
1972 *The Politics and Technology of Satellite Communications*. Lexington (MA): Lexington Books.

Gille, L.
1984 Croissance et Télécommunications. *Information, Communication et Croissance, le Bulletin de l'IDATE* 16:221–67.
1986 Growth and Telecommunications, in *Information, Telecommunications and Development*, 25–61. Geneva: ITU.

Gorove, S.
1984 Major Issues Arising from the Use of Geostationary Orbit, in *Regulation of Transnational Communications* [1984 Michigan Yearbook of International Legal Studies], edited by L.J. Anderson, 3–12. New York: Clark Boardman.

Hahn, R.W. and R.G. Noll
1983 Barriers to Implementing Tradable Air Pollution Permits: Problems of Regulatory Interactions. *Yale J. Reg.* 1(1):63–91.

Hampton, J.D.
1986 INTELSAT and Fiber Optic Cables: Challenge and Opportunity. Washington, D.C.: INTELSAT (Sept. 9).

Dale L. Hatfield Associates
1983 An Analysis of the INTELSAT Subsidy Issue (mimeo). Boulder (CO).

Hayek, F.A.
1968 Der Wettbewerb als Entdeckungsverfahren, in *Kieler Vorträge*, N.F. 56. Kiel (W. Ger.).

Walter Hinchman Associates, Inc.
1984 The Economics of International Satellite Communications (mimeo). Washington, D.C.

Hudson, H.E.
1983 The Role of Telecommunications in Development: A Synthesis of Current Research, in *Proceedings from the Tenth Annual Telecommunications Policy Research Conference*, edited by O.H. Gandy, Jr., P. Espinosa and J.A. Ordover, 291–308. Norwood (NJ): Ablex.
1984 *When the Telephone Reaches the Village: The Role of Telecommunications in Rural Development*. Norwood (NJ): Ablex.

Hudson, H.E. (ed.)
1985 *New Directions in Satellite Communications: Challenges for North and South*. Dedham (MA): Artech House.

Hudson, H.E., et al.
1979 *The Role of Telecommunications in Socio-Economic Development: A Review of the Literature with Guidelines for Further Investigations*. Washington, D.C.: Keewatin Communications.

INTELSAT
1986a *Bridging the Gap II: A Guide to Telecommunications and Development*. Washington, D.C.: INTELSAT.

1986b Project SHARE: Satellites for Health and Rural Education (paper delivered at the 17th Annual Conference of the International Institute of Communications). Washington, D.C.: INTELSAT.

International Telecommunication Union
1986 *Information, Telecommunications and Development*. Geneva: ITU.

Jansky, D.M. and M.C. Jeruchim
1983 *Communication Satellites in the Geostationary Orbit*. Dedham (MA): Artech House.

Jasentuliyana, N. and R. Chipman (eds.)
1984 *International Space Programmes and Policies: Proceedings of the Second United Nations Conference on the Peaceful Use of Outer Space (UNISPACE)* (held in Vienna in August 1982). New York: North-Holland.

Johnson, L.L.
1964 Joint Cost and Price Discrimination: The Case of Communications Satellites. *J. Bus.* 37(1):32–46.
1986 *Excess Capacity in International Telecommunications: Poor Traffic Forecasting or What?* Santa Monica (CA): RAND.
1987 *Issues in International Telecommunications: Government Regulation of Comsat*. Santa Monica (CA): RAND.

Jonscher, C.
1986 Telecommunications Liberalization in the United Kingdom, in *Marketplace for Telecommunications: Regulation and Deregulation in Industrialized Democracies*, edited by M.S. Snow, 153–72. New York: Longman.

Jussawalla, M.
1984 The Economic Implications of Satellite Technology and the Industrialization of Space. *Telecommunications Pol'y* 8(3):237–48.

Kahn, A.E.
1970–71 *The Economics of Regulation: Principles and Institutions*. 2 vols. New York: Wiley.

Kildow, J.T.
1973 *INTELSAT: Policy-maker's Dilemma*. Lexington (MA): Lexington Books.

Kinsley, M.E.
1976 *Outer Space and Inner Sanctums: Government, Business, and Satellite Communication*. New York: Wiley-Interscience.

Knieps, G., J. Müller and C.C. von Weizsäcker
1981 *Die Rolle des Wettbewerbs im Fernmeldebereich*. Baden-Baden (W. Ger.): Nomos.

Lamberton, D.M.
1986 Australian Regulatory Policy, in *Marketplace for Telecommunications: Regulation and Deregulation in Industrialized Democracies*, edited by M.S. Snow, 231–52. New York: Longman.

Ledyard, J.O.
1986 The Economics of the Space Station (Social Science Working Paper No. 617). Pasadena (CA): California Institute of Technology.

Leff, N.H.
1984a Externalities, Information Costs, and Social Benefit-Cost Analysis for Economic Development: An Example from Telecommunications. *Econ. Dev. & Cult. Change* 32(2):255–76.
1984b Social Benefit-Cost Analysis and Telecommunications in Developing Countries. *Infor. Econ. & Pol'y* 1(3):217–27.

Leive, D.M.
1984 Some Conflicting Trends in Satellite Telecommunications, in *Regulation of Transnational Communications* [1984 Michigan Yearbook of International Legal Studies], edited by L.J. Anderson, 73–82. New York: Clark Boardman.

Levin, H.J.
1965 Organization and Control of Communications Satellites. *U. Pa. L. Rev.* 113(1):315–57.
1970 Spectrum Allocation Without Market. *Am. Econ. Rev.* 60(2):209–18.
1971 *The Invisible Resource: Use and Regulation of the Radio Spectrum.* Baltimore (MD): Johns Hopkins University Press.
1980 *Fact and Fancy in Television Regulation: An Economic Study of Policy Alternatives.* New York: Russell Sage Foundation.
1984 The Political Economy of Orbit Spectrum Leasing, in *Regulation of Transnational Communications* [1984 Michigan Yearbook of International Legal Studies], edited by L.J. Anderson, 41–70. New York: Clark Boardman.
1985 Latecomer Cost Handicap in Satellite Communications. *Telecommunications Pol'y* 9(2):121–35.
1986 Emergent Markets for Orbit Spectrum Assignment: An Idea Whose Time Has Come (paper presented at the meetings of the American Economic Association, in New Orleans (LA)).

Levy, S.A.
1975 INTELSAT: Technology, Politics and the Transformation of a Regime. *Int'l Org.* 29(3):655–80.

Lewis, J.P.
1975 Postwar Economic Growth and Productivity in the Soviet Communications Industry. *Bell J. Econ.* 6(2):430–50.

Lovell, B.
1972 The Great Competition in Space. *Foreign Aff.* 51(1):124–38.

Lundberg, O.
1986 Competition and the Threat to INMARSAT. *Intermedia* 14(2):32–34.

Maddox, B.
1972 *Beyond Babel: New Directions in Communications.* New York: Simon & Schuster.

Maitland, D.
1984 *The Missing Link* (report of the Independent Commission for World Wide Telecommunications Development). Geneva: ITU.

Martinez, L.
1985 *Communications Satellites: Power Politics in Space.* Dedham (MA): Artech House.

McDougall, W.A.
1985 *... the Heavens and the Earth: A Political History of the Space Age.* New York: Basic Books.

Mestmäcker, E.-J.
1985 The Impact of Deregulation on Trade in Services Agreements and on Telecommunications Monopolies in Europe, in *Special Session of the World Telecommunication Forum, 1985: Law, Regulation, Standards of Global Communications*, edited by the ITU, 307–23. Geneva: ITU.

Mestmäcker, E.-J. (ed.)
1980 *Kommunikation ohne Monopole: Über Legitimation und Grenzen des Fernmeldemonopols.* Baden-Baden (W. Ger): Nomos.

Mickelson, S.
1969 Communications by Satellite. *Foreign Aff.* 68(1): 69–79.

Mitchell, B.M.
1978 Optimal Pricing of Local Telephone Service. *Am. Econ. Rev.* 68(4): 517–37.

Mowlana, H.
1977 Political and Social Implications of Communications Satellite Applications in Developed and Developing Countries, in *Economic and Policy Problems in Satellite Communications*, edited by J.N. Pelton and M.S. Snow, 124–42. New York: Praeger.

Nelson, R.W.
1975 Domestic Satellites, the FCC, and Competition in Domestic Telecommunications. *Land Econ.* 51(3):235–46.
1977 Domestic Satellite Communications: Economic Issues in a Regulated Industry Undergoing Technical Change, in *Economic and Policy Problems in Satellite Communications*, edited by J.N. Pelton and M.S. Snow, 31–61. New York: Praeger.

Neumann, K.-H. and T. Schnöring
1985 Das ISDN: Ein Problemfeld aus volkswirtschaftlicher und gesellschaftspolitischer Sicht (Diskussionsbeiträge zur Telekommunikationsforschung, No. 15). Koblenz: Deutsche Bundespost (Wissenschaftliches Institut für Kommunikationsdienste).

Neumann, K.-H., U. Schweizer and C.C. von Weizsäcker
1983 Welfare Analysis of Telecommunications Traffic in Germany, in *Public Sector Economics*, edited by J. Finsinger, 65–85. New York: St. Martin's Press.

Neumann, K.-H. and B. Wieland
1985 Competition and Social Objectives: The Case of West German Telecommunications (Diskussionsbeiträge zur Telekommunikationsforschung, No. 14). Koblenz: Deutsche Bundespost (Wissenschaftliches Institut für Kommunikationsdienste).

Noam, E.M.
1986a The Political Economy of ISDN: European Network Integration vs. American System Fragmentation (mimeo). New York.
1986b Telecommunications Policy on the Two Sides of the Atlantic: Divergence and Outlook, in *Marketplace for Telecommunications: Regulation and Deregulation in Industrialized Democracies*, edited by M.S. Snow, 255–74. New York: Longman.

Noll, R.G.
1983 The Political Foundations of Regulatory Policy. *Zeitschrift für die gesamte Staatswissenschaft* 139(3):377–404.
1985 Government Regulatory Behavior: A Multidisciplinary Survey and Synthesis, in *Regulatory Policy and the Social Sciences*, edited by R.G. Noll, 9–63. Berkeley (CA): University of California Press.
1986a Managing the Transition to Competition in Telecommunications (Studies in Industry Economics, Discussion Paper No. 135). Stanford (CA): Stanford University.
1986b The Political and Institutional Context of Communications Policy, in *Marketplace for Telecommunications: Regulation and Deregulation in Industrialized Democracies*, edited by M.S. Snow, 42–65. New York: Longman.

Noll, R.G. and B.M. Owen
1983 *The Political Economy of Deregulation: Interest Groups in the Regulatory Process*. Washington, D.C.: American Enterprise Institute for Public Policy Research.

Nosaka, K.
1984 The Earth Segment, in *The INTELSAT Global Satellite System*, edited by J.R. Alper and J.N. Pelton, 135–89. New York: American Institute of Aeronautics and Astronautics.

Olson, M.
1982 *The Rise and Decline of Nations: Economic Growth, Stagflation, and Social Rigidities*. New Haven (CT): Yale University Press.

Oslund, R.J.
1977 »Open Shores« to »Open Skies«: Sources and Directions of U.S. Satellite Policy, in *Economic and Policy Problems in Satellite Communications*, edited by J.N. Pelton and M.S. Snow, 143–99. New York: Praeger.

Pan American Satellite
1986 Information Required for Non-Technical Consultation of the Pan American Satellite System Under Article XIV(d) of the Intelsat Agreement (mimeo). Greenwich (CT).

Park, R.E. and B.M. Mitchell
1986 *Optimal Peak-Load Pricing for Local Telephone Calls*. Santa Monica (CA): RAND.

Pelton, J.N.
1974 *Global Communication Satellite Policy: INTELSAT, Politics, and Functionalism*. Mt. Airy (MD): Lomond Books.
1981 *Global Talk*. Alphen an den Rijn (Neth.): Sijthoff & Noordhoff.
1982 The Pacific, Telecommunications and Socio-Economic Development, in *Pacific Telecommunications Conference Proceedings 1982*, edited by D.J. Wedemeyer, 143–49. Honolulu (Hawaii): Pacific Telecommunications Council.

Pelton, J.N. and J.R. Alper
1984 The Next Hundred Years: INTELSAT and the Postinformation Society, in *The INTELSAT Global System*, edited by J.R. Alper and J.N. Pelton, 363–78. New York: American Institute of Aeronautics and Astronautics.

Pelton, J.N. and D.A. Demac
1986 *Telecommunications and Development*. Washington, D.C.: INTELSAT.

Pelton, J.N. and M.S. Snow (eds.)
1977 *Economic and Policy Problems in Satellite Communications*. New York: Praeger.

Perras, M.
1984 The Development of New INTELSAT Services, in *The INTELSAT Global System*, edited by J.R. Alper and J.N. Pelton, 255–68. New York: American Institute of Aeronautics and Astronautics.

Pierce, W. and N. Jequier
1983 *Telecommunications for Development*. Geneva: ITU.

Podraczky, E. and J.N. Pelton
1984 INTELSAT Satellites, in *The INTELSAT Global Satellite System*, edited by J.R. Alper and J.N. Pelton, 95–133. New York: American Institute of Aeronautics and Astronautics.

Potamitis, S.
1986 Competition in International Satellite Telecommunications Services. *U. Toronto Fac. L. Rev.* 44(1):33–56.

Pulley, L.B. and Y.M. Braunstein
1984 Scope and Scale Augmenting Technological Change: An Application in the Information Sector, in *Communication and Information Economics: New Perspectives*, edited by M. Jussawalla and H. Ebenfield, 105–18. New York: North-Holland.

Rahim, S.A.
1983 Telecommunications Technology and Policy: The Case of Satellite Communications. *Keio Com. Rev.* 4:51–58.

Rahim, S.A. and D.J. Wedemeyer (eds.)
1983 *Telecom Pacific*. Honolulu (Hawaii): Pacific Telecommunications Council.

Ramsey, F.
1927 A Contribution to the Theory of Taxation. *Econ. J.* 37(1):47–61.

Rea, J.D. and G.M. Lage
1978 Estimates of Demand Elasticities for International Telecommunications Service. *J. Indus. Econ.* 26(4):363–81.

Research Institute for Telecommunications and Economics (Japan)
1983 A Study on the Pacific Regional Satellite Communications System (paper presented at the Pacific Telecommunications Conference, in Honolulu (Hawaii)).

Riegel, O.W.
1971 Communications by Satellite: The Political Barriers. *Q. Rev. Econ. & Bus.* 11(4):23–35.

Rittig, G.
1977 *Gemeinwirtschaftsprinzip und Preisbildung unter volkswirtschaftlichen Gesichtspunkten* (Schriftenreihe Gemeinwirtschaft 25). Frankfurt am Main (W. Ger.).

Rutkowski, A.M.
1984 Integrated Services Digital Network: Issues and Options for the World's Future Communications Systems, in *Regulation of Transnational Communications* [1984 Michigan Yearbook of International Legal Studies], edited by L.J.Anderson, 243–70. New York: Clark Boardman.

Sarreals, C.L.
1986 International Telecommunications Satellite Services: The Spirit of Cooperation versus the Battle for Competition. *Jurimetrics J.* 26(3):367–404.

Saunders, R.J.
1983 Telecommunications in the Developing World: Investment Decision and Performance Monitoring. *Telecommunications Pol'y* 7(7):277–84.

Saunders, R.J., J.J. Warford and B. Wellenius
1983 *Telecommunications and Economic Development*. Baltimore (MD): Johns Hopkins University Press.

Schnicke, W.R.
1984 INTELSAT System Planning, in *The INTELSAT Global Satellite System*, edited by J.R. Alper and J.N. Pelton, 311–30. New York: American Institute of Aeronautics and Astronautics.

Schwartz, H.
1965 Governmentally Appointed Directors in a Private Corporation – The Communications Satellite Act of 1962. *Harv. L. Rev.* 79:350–64.
1967 Comsat, the Carriers, and the Earth Stations: Some Problems with »Melding Variegated Interests«. *Yale L.J.* 76:441–84.

Shapley, D.
1985 *The Seventh Continent: Antarctica in a Resource Age*. Washington, D.C.: Resources for the Future.

Sharkey, W.M.
1982 *The Theory of Natural Monopoly*. New York: Cambridge University Press.

Shepherd, W.G.
1983 Concepts of Competition and Efficient Policy in the Telecommunications Sector, in *Telecommunications Regulation Today and Tomorrow*, edited by E.M. Noam, 79–120. New York: Harcourt Brace Jovanovich.
1984 »Contestability« vs. Competition. *Am. Econ. Rev.* 74(4):572–87.

Shillinglaw, T.L.
1970 The Soviet Union and International Satellite Telecommunications. *Stan. J. Int'l Stud.* 5:199–226.

Snow, M.S.
1975 Investment Cost Minimization for Communications Satellite Capacity: Refinement and Application of the Chenery-Manne-Srinivasan Model. *Bell J. Econ.* 6(2):621–43.
1976 *International Commercial Satellite Communications: Economic and Political Issues of the First Decade of INTELSAT.* New York: Praeger.
1977a INTELSAT and Domestic Satellite Systems in the United States and Abroad (paper presented at the Fifth Annual Telecommunications Policy Research Conference, in Airlie (VA)) (mimeo).
1977b Price Discrimination and Economies of Scale in International Satellite Communications, in *Economic and Policy Problems in Satellite Communications*, edited by J.N. Pelton and M.S. Snow, 5–30. New York: Praeger.
1978 Problems in Regulating New Information Technology at the National and International Levels: The Case of Communications Satellites, in *Information Societies: Comparing the Japanese and American Experiences*, edited by A.S. Edelstein, J.E. Bowes and S.M. Harsel, 55–67. Seattle (WA): University of Washington Press.
1979 Some Effects of Cost-Deaveraging by INTELSAT (consultant's report to the National Telecommunications and Information Administration, U.S. Department of Commerce) (mimeo).
1980 INTELSAT: An International Example. *J. Com.* 30(2):147–56.
1982 Economic, Financial, and Institutional Aspects of a Communications Satellite System for the Pacific Basin (consultant's report to the U.S. National Aeronautics and Space Administration), published as part of *Satellite Communications in the Pacific Islands.* Washington, D.C.: NASA
1984 The State as Stopgap: Social Economy and Sustainability of Monopoly in the Telecommunications Sector (mimeo), forthcoming in *Rev. Soc. Econ.*
1985a Arguments For and Against Competition in International Satellite Facilities and Services: A U.S. Perspective. *J. Com.* 35(3):51–79.
1985b Regulation to Deregulation: The Telecommunications Sector and Industrialization, with examples from the Pacific Rim and Basin. *Telecommunications Pol'y* 9(4):281–90.
1986a Communications Policy in Seven Developed Countries: Introduction, Background, and Conclusions, in *Marketplace for Telecommunications: Regulation and Deregulation in Industrialized Democracies*, edited by M.S. Snow, 3–19. New York: Longman.
1986b Global Commercial Satellite Communication: Twenty Years of Peaceful Cooperative Usage (paper presented at the Pacific Peace Seminar, in Honolulu (Hawaii)).
1986c Regulating Telecommunications, Information and the Media: An Agenda for Future Comparative Research, in *Marketplace for Telecommunications: Regulation and Deregulation in Industrialized Democracies*, edited by M.S. Snow, 275–94. New York: Longman.
1987a Economics of Satellite Communications, in *Satellites International*, edited by J. Howland and J.N. Pelton, vol. 1 (forthcoming). London: Macmillan.
1987b Natural Monopoly in INTELSAT: Cost Estimation and Policy Implications for the Separate Systems Issue. *Telematics and Informatics* (forthcoming).

Snow, M.S. (ed.)
1986 *Marketplace for Telecommunications: Regulation and Deregulation in Industrialized Democracies.* New York: Longman.

Snow, M.S. and M. Jussawalla
1986 *Telecommunication Economics and International Regulatory Policy: An Annotated Bibliography.* Westport (CT): Greenwood Press.

Spence, A.M.
1983 Contestable Markets and the Theory of Industry Structure; A Review Article. *J. Econ. Lit.* 21(3):981–90.

Srinivasan, T.N.
1967 Geometric Rate of Growth of Demand, in *Investments for Capacity Expansion*, edited by A.S. Manne. Cambridge (MA): MIT Press.

Stanley, K.B.
1976 Pricing of Satellite Services in the International Telecommunications Industry, in *New Dimensions in Public Utility Pricing*, edited by H.M. Trebing, 403–22. East Lansing (MI): Michigan State University Press.
1977 Economic Issues in International Telecommunications: A Public Policy Dilemma, in *Economic and Policy Problems in Satellite Communications*, edited by J.N. Pelton and M.S. Snow, 62–88. New York: Praeger.

Sueyoshi, T. and P.C. Anselmo
1986 The Evans and Heckman Subadditivity Test: Comment. *Am. Econ. Rev.* 76(4):854–55.

Taylor, L.D.
1980 *Telecommunications Demand: A Survey and Critique*. Cambridge (MA): Ballinger.
1983 Problems and Issues in Modeling Telecommunications Demand, in *Economic Analysis of Telecommunications: Theory and Applications*, edited by L. Courville, A. de Fontenay and R. Dobell, 181–98. New York: North-Holland.

Thiemeyer, T.
1970 *Gemeinwirtschaft als Ordnungsprinzip: Grundlage einer Theorie gemeinnütziger Unternehmen*. Berlin (West): Duncker & Humblot.
1983 Deregulation in the Perspective of the German *Gemeinwirtschaftslehre. Zeitschrift für die gesamte Staatswissenschaft* 139(3):405–18.

Tucker, E.
1985 Down-to-Earth Lobbying About Communication in Space. *Wash. Post Nat'l Weekly Ed.* 2(22):14.

Tudge, D.D.
1987 Global Aspects of ISDN. Washington, D.C.: INTELSAT (Jan. 19).

U.S. Congress
1962 Communications Satellite Act of 1962, 47 U.S.C. §§ 701 et seq. (1982).

U.S. Department of Commerce
1984 Letter from Malcolm Baldridge, Secretary of Commerce, and George P. Shultz, Secretary of State, to Mark S. Fowler, Chairman, Federal Communications Commission (Nov. 28).

U.S. Department of State
1964 Agreement Establishing Interim Arrangements for a Global Commercial Communications Satellite System, Aug. 20, T.I.A.S. No. 5646.
1971 Agreement Relating to the International Telecommunications Satellite Organization, Aug. 20, T.I.A.S. No. 7532.

U.S. Federal Communications Commission
1985 Establishment of Satellite Systems Providing International Communications, 101 F.C.C.2d 1046.
1986 Memorandum Opinion and Order, FCC 86–144, 36563 (Apr. 3).

U.S. House of Representatives
1981 Majority Staff of the Subcommittee on Telecommunications, Consumer Protection, and Finance, Committee on Energy and Commerce, Telecommunications in Transition: The Status of Competition in the Telecommunications Industry. 97th Cong., 1st Sess. Washington, D.C.: Government Printing Office.

U.S. National Commission on Space
1986 *Pioneering the Space Frontier: An Exciting Vision of Our Next Fifty Years in Space*. New York: Bantam Books.

U.S. Office of Technology Assessment
1984 *International Cooperation and Competition in Civilian Space Activities*. Washington, D.C.

U.S. President
1986 Commission on the Space Shuttle Challenger Accident: Final Report. Washington, D.C. (June 6).
1984 Presidential Determination No. 85–2 (Nov. 28).

U.S. Senate
1983 Committee on Commerce, Science, and Transportation, Long-Range Goals in International Telecommunications and Information: An Outline for United States Policy, 98th Cong., 1st Sess. Washington D.C.: Government Printing Office.

U.S. Senior Interagency Group on International Communication and Information Policy
1985 A White Paper on New International Satellite Systems (mimeo). Washington, D.C.

Voge, J.P.
1986 Survey of French Regulatory Policy, in *Marketplace for Telecommunications: Regulation and Deregulation in Industrialized Democracies*, edited by M.S. Snow, 106–30. New York: Longman.

von Mises, L.
1981 *Socialism: An Economic and Sociological Analysis*. Translated by J. Kahane from the 1932 (second) German edition. Indianapolis (IN): Liberty Classics.

von Weizsäcker, C.C.
1986 Free Entry into Telecommunications?, in *Marketplace for Telecommunications: Regulation and Deregulation in Industrialized Democracies*, edited by M.S. Snow, 20–41. New York: Longman.

Wenders, J.T.
1987 *The Economics of Telecommunications: Theory and Policy*. Cambridge (MA): Ballinger.

Withers, D.J. and H.J. Weiss
1984 INTELSAT and the ITU, in *The INTELSAT Global Satellite System*, edited by J.R. Alper and J.N. Pelton, 269–310. New York: American Institute of Aeronautics and Astronautics.

Zajac, E.E.
1978 *Fairness of Efficiency: An Introduction to Public Utility Pricing*. Cambridge (MA): Ballinger.

No Author
1984 Separate International Communications Satellite Systems: Three Documents. *Telematics and Informatics* 1(4):451–55.

Appendix

A White Paper on New International Satellite Systems

Senior Interagency Group on International Communication and Information Policy

William Schneider, Jr.
Under Secretary for Security Assistance, Science, and Technology
U.S. Department of State

David J. Markey
Assistant Secretary for Communications and Information
U.S. Department of Commerce

February 1985

Index

Introduction

Since 1983, several U.S. firms have filed applications with the Federal Communications Commission (FCC) to establish international communications satellite systems in addition to the global system owned by the International Telecommunications Satellite Organization (INTELSAT). Orion Satellite Corporation, International Satellite, Inc. (ISI), and Cygnus Corporation propose new transatlantic communications systems, and RCA American Communications, Inc. (RCA) has applied to use capacity on a U.S. domestic satellite to provide international service. Pan American Satellite Corporation (PanAmSat) proposes to establish a system which would serve Latin America. In addition to existing and planned regional satellite systems independent of INTELSAT, other transoceanic satellite systems are under consideration abroad. Approved and proposed transatlantic submarine cable communications facilities, many of which are actually or potentially competitive with INTELSAT, are pending as well.

Focus of Report

The filing of U.S.-based satellite system applications with the FCC prompted action by the Executive branch, which has special responsibilities in this field under the Communications Satellite Act of 1962, as amended (47 U.S.C. 701 et seq.), including the responsibility to determine whether additional U.S. international satellite systems are »required in the national interest.« The Senior Interagency Group on International Communication and Information Policy (SIG) reviewed U.S. international satellite policy to determine whether, and under what conditions, authorizing satellite systems and services in addition to INTELSAT would be: (a) consistent with prevailing U.S. law, practice, and international treaty obligations; (b) compatible with sound foreign policy and telecommunications policy goals; and, (c) in the U.S. national interest.[1]
The Executive agencies represented on the SIG undertook a study and reached a unanimous position in favor of new entry, subject to certain limitations. A recommendation subsequently was made to the President by the Secretaries of State and Commerce. The President determined on November 28, 1984, that international satellite systems separate from INTELSAT were required in the U.S. national interest, subject to certain conditions. Specific criteria relating to the President's determination were then forwarded to the FCC by the Secretaries of Commerce and State jointly. See Appendixes A and B.
This report provides background information regarding the President's determination, and it also provides information on important regulatory and other parallel measures which are desirable to ensure that the Executive branch's fundamental policy goal – an

1 The SIG is composed of representatives of the Departments of State, Justice, Defense, and Commerce; the Offices of Management and Budget, Science and Technology Policy, Policy Development, and the U.S. Trade Representative; the National Security Council; the Central Intelligence Agency; the U.S. Information Agency (USIA); the Board for International Broadcasting; the Agency for International Development; and the National Aeronautics and Space Administration. Commerce and State co-chair the SIG, and USIA serves as vice chair.

efficient and responsive international communications environment – is achieved. The discussion here focuses on the major communications and information policy issues raised by the applications before the FCC. It addresses commercial, trade, and legal matters, and also examines major U.S. foreign policy interests and concerns.
This report does not seek to resolve all of the questions that have been raised regarding new international satellite systems nor to direct action by the FCC on specific pending applications. It does, however, consolidate much of the extensive analysis that has been undertaken by the Executive branch and sets forth the requirements applicable to any system the FCC may eventually authorize.
The Executive branch has concluded, in brief, that it is technically feasible, economically desirable, and in the national interest to allow new entry by U.S. firms into the international satellite field. Customers should be afforded both the new service options and the benefits of competition among customized service providers that new entry promises. This can be accomplished, moreover, while maintaining the technical integrity of the INTELSAT global system and avoiding significant economic harm to that system. U.S. foreign policy, and international communications and information policy, require a continued strong national commitment to INTELSAT as »a single global commercial telecommunications satellite system as part of an improved global telecommunications network.«[2] But our national commitment to INTELSAT and other important goals can be accommodated, provided that new international satellite systems and services are authorized and regulated along the lines discussed in this report.
Specifically, this report concludes that –

(a) Additional international satellite facilities should be permitted by the FCC, provided they satisfy conventional regulatory requirements, but the new entrants must be restricted to providing customized services, as defined in this report. When one or more authorities abroad authorizes use of such new systems, the United States with those authorities will enter into consultation procedures with INTELSAT under Article XIV(d) of the INTELSAT Agreement. Construction permits may be issued at the conclusion of regulatory proceedings to those applicants meeting the public interest requirements of the Communications Act. Final licenses and authorizations should not be issued, however, until after INTELSAT consultation is completed.

(b) The FCC should examine allowing U.S. carriers and users in addition to the Communications Satellite Corporation (Comsat) to have cost-based access to the INTELSAT space segment for customized services. This matter can be pursued on a parallel track, as the pending applications are being processed, however, and does not constitute a condition to FCC action on these applications.

(c) The United States should, and will, maintain its full commitment to INTELSAT, while permitting technology-driven competition in this important sector to envolve.

2 Preamble, Agreement Relating to the International Telecommunications Satellite Organization »INTELSAT«, TIAS 7532, 23 UST 3813, 3814 (1973).

I. *The International Communications Marketplace Today*

Industry Participants

International communications today constitutes one of the most rapidly growing parts of the overall telecommunications industry, and the services involved are critical to U.S. trade, national defense, foreign policy, and international investment. The services involved traditionally have been categorized as »voice« or »record,« »private line« or »public-switched,« and, historically, the American Telephone and Telegraph Company (AT&T) has handled most of the international voice traffic. Six major international record carriers (IRCs) – ITT Worldcom, RCA Communications, MCI International, TRT Communications, Western Union, and FTC Communications – currently share the telex and telegraph components of the $2.8 billion a year international communications market. AT&T and the IRCs competitively offer international private line services, generally used by major corporate and Government users for data and voice communication.[3]

There are two principal international transmission modes: submarine cables and communications satellite facilities. The submarine cables which provide U.S. international service are owned collectively by AT&T, the IRCs, and their foreign correspondents.[4] Seven transatlantic cables now terminate in the United States and an eighth, 38,000-circuit, fiber optic cable has been approved by the FCC.[5] U.S. international satellite circuits are provided by Comsat, which has functioned as a »carrier's carrier« and holds a 23 percent interest in INTELSAT, the 109-nation organization that owns and manages the global satellite system.[6] Comsat's investment share is adjusted annually to reflect U.S. use of the INTELSAT system. The U.S. earth stations used to provide access to the satellites are now collectively owned by Comsat, AT&T, and the IRCs, although the FCC recently made changes in this regard.[7]

3 *See* Overseas Communications Services, 92 F.C.C.2d 641 (1982). *See also* WUI, Inc. v. FCC, 673 F.2d 539 (D.C. Cir. 1982) (and citations therein); TAT-5, 13 F.C.C.2d 263 (1968). *See generally* Comsat Rate Case, 56 F.C.C.2d 1101 (1975), *aff'd*, 611 F.2d 883 (D.C. Cir. 1977); Comsat Study, 77 F.C.C.2d 564 (1982); Comsat Structure Decision, 52 P.&F. Radio Reg.2d 153 (1982); Schwartz, *Comsat, the Carriers, and the Earth Stations: Some Problems With »Melding Variegated Interests«*, 76 YALE L.J. 441 (1967).

4 Such correspondents typically consist of government-owned (or, in the case of Britain and Japan, »privatized«) postal, telephone, and telegraph administrations (PTTs) that both provide and regulate domestic and international communications services.

5 *See* Applications of AT&T et al. (File No. ITC 84-072), FCC Mimeo 84-240 (June 8, 1984). In addition, two sets of applications to install noncarrier-owned transatlantic cables are now before the FCC.

6 The FCC has recently increased the range of »authorized users« to whom Comsat may provide services. Authorized User II Policy (CC Dkt 80-170), FCC Mimeo 84-633 (Dec. 19, 1984).

7 *See* Earth Station Ownership, 90 F.C.C.2d 1958 (1982); Modification of Earth Station Policies (CC Docket No. 82-540), FCC Mimeo 84-605 (released Dec. 18, 1984). *See generally Twenty-First Comsat Report to the President and the Congress* at pp. 2–3, 6–7 (Sept. 17, 1984).

U.S. Policy Goals

The international communications and information policy goals of the United States are the following:

- To enhance the free flow of information and ideas among nations;
- To promote harmonious international relations and contribute to world peace and understanding through communications;
- To promote, in cooperation with other nations, the development of efficient, innovative, and cost-effective international communications services responsive to the needs of users and supportive of the expanding requirements of commerce and trade;
- To ensure the continued technological and economic strength and leadership of the United States in the communications, information, and aerospace fields;
- To expand U.S. private sector investment and involvement in civil space and related activities;
- To promote expanded international trade and to ensure opportunities to U.S. firms to participate in such trade;
- To promote the continuing evolution of an international configuration of communications services that can meet the needs of all nations of the world, with attention toward providing such services to developing nations;
- To ensure efficient utilization of the geostationary orbit and the electromagnetic radio frequency spectrum;
- To promote competition and reliance on market mechanisms, as feasible, and to foster cost-based pricing, quality service, and more efficient use of resources; and,
- To ensure the needs of national defense, security, and emergency preparedness are satisfactorily met.

These basic policy goals are mutually supportive. There is a continuing need to review and assess their requirements. Satisfying all of these goals to the maximum extent possible requires striking a reasonable balance.

Evolution in International Communications

The U.S. international communications business has experienced sustained rapid growth and fundamental regulatory changes in recent years as technology has advanced, demand has grown, and the level of competition has increased. International telephone calls increased more than 15 percent in 1984, for example, producing revenues of over $2 billion. The IRCs in 1984 are expected to report operating revenues exceeding $650 million, up from $617 million in 1983. Overseas circuits used by the IRCs grew to 2,874 in 1983, an 8.2 percent increase over 1982 levels, despite depressed economic conditions worldwide. Changes in FCC »gateway« and related regulations have permitted the IRCs to offer an expanding customer base improved and more responsive services. Under the Record Carrier Competition Act of 1981 (Public Law No. 97–130), Western Union was permitted to reenter the international record communications business. GTE Sprint and

MCI, important U.S. competitive carriers domestically, have entered the international field and will both begin providing international telephone service in 1985.[8]

Domestically, the continuing advent of International Direct Distance Dialing (IDDD), which enables subscribers to dial a growing number of nations without operator assistance, has facilitated international telephone calling. By 1983, IDDD capability existed in 86 locations around the world, and about 60 percent of U.S. telephone subscribers enjoyed this capability. Continued installation of advanced electronic switching is expected to boost U.S. IDDD penetration to about 67 percent by the end of 1984. This should have a positive effect on the overall volume of public-switched message traffic.

Steadily growing demand for conventional international communications services is reflected in other statistics as well. Comsat's World Systems Division, for instance, obtains circuit capacity from INTELSAT and provides that capacity to other U.S. international carriers for telephone, data, telex, and facsimile services. The volume of Comsat's communications business through INTELSAT increased 6 percent between 1982 and 1983, again despite a worldwide economic downturn, and notwithstanding the loading of a new transatlantic cable (TAT-7), which shared traffic growth with the satellite network. In 1983, Comsat's regulated satellite services accounted for most of the firm's revenues of $440 million and net income of $50 million. The FCC in 1982 authorized Comsat to retail certain services (such as television transmission service) directly to end-users, altering its traditional policy of restricting Comsat to serving as a »carrier's carrier.« The FCC's legal authority to do so was sustained on appeal, although its decision was remanded for further consideration.[9] Recently, the FCC again ruled that expanded retail activities by Comsat are in the public interest and this action should have a beneficial effect on the volume of Comsat's business.[10]

The INTELSAT system and the number of facilities which access INTELSAT's satellites have expanded rapidly. INTELSAT's 15 satellites today serve 173 countries, territories, and possessions directly or indirectly, and the organization leases satellite capacity to 26 nations for domestic services. Nineteen new earth stations and 39 new international communications antennas were added in 1982 alone. As of November 1984, there were 198 INTELSAT earth station sites and 293 international antennas in 157 countries, dependencies, and areas of other special sovereignty.

INTELSAT now handles about two-thirds of the world's transoceanic telecommunications traffic and most international television transmissions. Demand for full-time voice, record, and data services for INTELSAT grew by 18 percent in 1982; these services accounted for about 86 percent of the total satellite utilization revenue received by INTELSAT that year. The most recently published INTELSAT annual report states

8 *See generally* 1985 U.S. INDUSTRIAL OUTLOOK, ch. 31 (U.S. Department of Commerce, 1985). In contrast to the domestic sector, where record communications are marginally significant, differences in language and time zones make international record communications services commercially critical.

9 *See* Modification of Authorized User Policy, 90 F.C.C.2d 1934 (1982), *rev'd sub nom.* ITT Worldcom v. FCC, 725 F.2d 732 (D.C. Cir. 1984).

10 *See* note 6, *supra*.

that INTELSAT expects continued strong growth of 15 percent annually on an expanded base of conventional international traffic over the 1988–2000 time period.[11] With the growth of the INTELSAT system, circuit charges have steadily declined. The annual charge for a 1965 INTELSAT I »Early Bird« half-circuit, for example, was $32,000, while the 1982–83 charge for an equivalent, though technically superior, half-circuit was $4,680. There is disagreement, however, over whether the substantial INTELSAT charge reductions over the past decades have been fully reflected in the prices which Comsat has charged U.S. international carriers or the prices which those carriers have charged their customers. At present, end-user prices for many international satellite services both here and abroad typically are between two to ten times INTELSAT's charges.[12] U.S. international communications costs, moreover, often are very substantially above those for comparable domestic service.

INTELSAT has continued to grow and to prosper in an increasingly competitive international communications environment. Since 1981, the FCC has sanctioned certain international communications services using U.S. domestic satellite systems. At present, U.S. and Canadian satellites are used to provide certain services throughout North America and the Caribbean.[13] Additional proposals for such transborder satellite service will be the object of consultations with INTELSAT. INTELSAT recently accelerated its plans and now offers a number of international communications services aimed particularly at meeting specialized and sophisticated business community needs.

Significance to Industry and Government

International communications services constitute an essential component of international trade today. Efficient and effective international communications are necessary to international finance, to facilitate the production and shipment of goods, and to manage U.S. off-shore operations, assets, and investments.[14] International communications are also critical to the continued development of U.S. trade in services, which exceeded $40 billion in 1982.[15] International communications, moreover, play a central role in

11 1983 *INTELSAT Annual Report* at pp. 10, 17 (Mar. 1984). There are indications that INTELSAT's rate of growth declined in 1983 and 1984, but official statistics have not yet been published. Similarly, it has been suggested that the mix of traffic also has changed. The statistics here are from the most recent official reports available to the public.

12 *See* Price of International Satellite Service: Comsat vs. INTELSAT (NTIA Rep. No. 83-122); Statement of then-INTELSAT Director General-elect Colino Before the Senate Subcommittee on Arms Control, Oceans, International Operations, and Environment, 98th Cong., 2d Sess., at p. 33 and Appendix 5 (Oct. 19, 1983); Colino, *The INTELSAT System: An Overview*, in THE INTELSAT GLOBAL SATELLITE SYSTEM (S. Alper & J. Pelton, eds.) (AIAA, 1984).

13 *See* Transborder Satellite Video Services, 88 F.C.C.2d 258 (1981); FCC Common Carrier Bureau Order No. 6119 (1983).

14 *See generally* Bryant & Krause, WORLD ECONOMIC INTERDEPENDECE IN SETTING NATIONAL PRIORITIES: AGENDA FOR THE 1980s (J.A. Pechman, ed., Brookings Inst., 1980) at pp. 71, 74; Saunders, Warford & Wellenius, TELECOMMUNICATIONS AND ECONOMIC DEVELOPMENT (World Bank, 1983) at pp. 100–02.

15 *See, e.g.*, LONG-RANGE GOALS IN INTERNATIONAL TELECOMMUNICATIONS AND INFORMATION at p. 155 et seq. (NTIA, 1983) (reprinted as Senate Commerce Committee Print No. 98-22, 98th Cong., 1st Sess.); 1985 U.S. INDUSTRIAL OUTLOOK at p. 38 (U.S. Department of Commerce, 1985).

facilitating the further economic development of less developed nations, thus permitting these countries to participate more fully in the world economy and contributing to peace, stability, and greater understanding.

Space communications is a major part of the aerospace industry, one of the world economy's most important »high-tech« or »sunrise« sectors, and an area where the excellence of U.S. manufacturing techniques and high technologies is reflected in the preeminence of the U.S. aerospace industry. U.S. aerospace trade is forecast to accelerate in 1985 as both exports and imports reach record highs. Aerospace exports should climb to a projected $18.9 billion, while imports will rise to $5.0 billion. The resultant trade surplus of $13.9 billion will be more than 30 percent above the level recorded in 1984. Total U.S. aerospace employment will rise an estimated 4 percent in 1985 to 739,000, with an estimated gain of 7 percent in the number of production workers.[16]

Existing Policies and Objectives

U.S. policy regarding international public-switched message services via satellite has centered on the 1962 Satellite Act and associated Executive Orders for more than 20 years. The 1962 Act authorized the establishment of Comsat and franchised it to serve as the U.S. private sector commercial participant in the INTELSAT system.

The established foreign and domestic policies of the U.S. Government in this area seek to further the basic goals which are outlined above. These policies include:

- Adhering to the requirements and provisions of the Communications Act of 1934, as amended (47 U.S.C. 151 et seq.) and the 1962 Satellite Act, as amended (47 U.S.C. 701 et seq.);
- Complying with the terms of the INTELSAT Agreement (TIAS 7532) and all the privileges and obligations the Agreement provides its Parties and Signatories;
- Supporting INTELSAT as »a single global commercial telecommunications satellite system as part of an improved global telecommunications network« (Preamble, INTELSAT Agreement), and as a key element providing all countries of the world access to global communications services;
- Concurring in the development, separate from INTELSAT, of customized, regional, and transborder satellite services where technical or economic consultation, or both, is accomplished as required under the terms of the INTELSAT Agreement and such systems are consistent with the Agreement;
- Pursuing a nondiscriminatory satellite launch policy;
- Adopting domestic communications policies which emphasize reduced Government regulation, wherever feasible, and increased reliance on market forces in the provision of communications and information services;
- Advocating and adopting international communications policies which stress reliance on free enterprise, competition, and free trade, wherever feasible, with full recognition that provision of international communications and information services involves the joint undertakings among sovereign nations requiring mutually acceptable agreements to accommodate differing national policies;

16 *See* 1985 U.S. Industrial Outlook, at p. 37–39 (U.S. Department of Commerce, 1985).

- Supporting and fostering the development of a diversity of international communications technologies and modes, including fixed, mobile, and broadcast satellite, microware, terrestrial and undersea cable, and optical fiber;
- Supporting and undertaking bilateral consultations and agreements, as well as multilateral deliberations in appropriate international forums, to ensure order and cooperation in the evolution of international communications and information services.

Already Competitive Environment

It is important to bear in mind that the pending proposals to establish U.S. international satellite systems separate from INTELSAT represent only possible incremental – not fundamental – competitive change in an international communications environment which is already characterized by some competition. The present proceeding thus does not pose choices directly comparable to those presented in 1968 when the FCC approved domestic, facilities-based competition by companies, such as MCI, with the dominant long-distance carrier, AT&T, or in 1970, when the FCC considered adoption of an »open skies« policy regarding proposed U.S. domestic satellite systems.[17] Despite significant regulation of the international communications industry both here and abroad, there nevertheless is competition between the extensive submarine cable facilities owned by terrestrial carriers and the satellite and earth station facilities owned by Comsat and INTELSAT. This competition stands to increase when high-capacity fiber optic cables – both carrier-owned and, perhaps, noncarrier-owned as well – become operational, especially if the traditional »balanced loading« rules governing the apportionment of traffic are changed.
There has also been competition among satellite systems for several years. As noted, U.S. and Canadian domestic satellite systems, for instance, have been authorized to handle traffic that is technically »international« – involving Canada, the United States, and Caribbean nations and locations. »Domestic-overseas« traffic to Alaska, Hawaii, and U.S. possessions which previously transited Comsat and INTELSAT facilities is now handled by U.S. domestic satellite systems.[18]
Nor is this emerging actual and potential competition limited to the Western Hemisphere by any means. Regional satellite systems operate in Southeast Asia and Europe and are planned for the Middle East and, perhaps, Africa as well. Several European administrations also plan soon to deploy »domestic« satellite systems which are capable of providing transatlantic service. The »footprints« of the planned British and French domestic satellite systems, for example, cover much of the eastern half of the United States and Canada. Extensive submarine cable facilities, moreover, are under construction in the Mediterranean, Indian Ocean, and Pacific region. There is no evidence, in this regard, that these new communications systems have had any adverse impact on the technical or economic integrity of the INTELSAT global system.

17 *See generally* Washington Util. & Transp. Comm'n v. FCC, 513 F.2d 1142 (9th Cir.), *cert. denied*, 423 U.S. 836 (1975); Network Project, Inc. v. FCC, 511 F.2d 786 (D.C. 1975).
18 *See, e.g.*, Colino, *International Cooperation Between Communications Satellite Systems: An Overview of Current Practices and Future Prospects*, 5 J. SPACE L. 65, 92 (1977).

II. *Institutional Limits on Competition*

The United States since the early 1970s consistently has sought to reduce outmoded communications regulation and to eliminate unnecessary barriers to competition chiefly domestically, but internationally as well. Important changes and regulatory reforms have been accomplished.[19] All recognize, however, that achieving a regulation-free international communications environment is not foreseeable at this time. There will remain significant U.S. limitations on competition in international communications as well as limits imposed by communications administrations abroad. Understanding some of these limits on potential competition is important to addressing the issues presented by the satellite applications pending before the FCC and reinforces our assessment that these applications imply continued evolutionary development, not radical or disruptive change.

Regulatory Constraints

There are, to begin with, a number of statutory requirements and limitations which bear on the level and intensity of potential competition in the international communications field. To enter the international communications satellite business, U.S. firms require FCC permission under title III of the 1934 Communications Act, provisions of title II of that Act (for would-be common carrier entrants), as well as provisions of the 1962 Satellite Act. The FCC is required by law to make an affirmative »public interest« finding prior to issuing construction permits and licenses to use the radio frequency spectrum.[20] Considerable regulatory review of proposed systems typically is entailed. It is also relevant in this regard to note that given spectrum use limitations and international procedures governing the use of the geostationary orbital resource, there are significant technical constraints on possible entry into international satellite communications.[21]

Entrants proposing to operate on a common carrier basis are subject to many provisions of title II of the 1934 Communications Act (e.g., 47 U.S.C. 214). Under title II, the FCC must generally find that the public interest, convenience, and necessity will be furthered by approving an additional international common carrier facility. Existing common carriers, moreover, must generally receive permission to make use of new facilities. As with other regulatory agencies, the FCC is required to weigh competitive factors when it functions as a »gatekeeper« with respect to common carrier communications.[22] Under present law, however, the FCC may not legally authorize new common

19 *See* Computer and Communications Industry Assoc. v. FCC, 693 F.2d 198 (D.C. Cir. 1982); Detariffing International Enhanced Services, FCC Docket Nos. RM-4435, CC 83-1230 (1983).

20 *See, e.g.*, Telocator Network of America v. FCC, 691 F.2d 525, 548 (D.C. Cir. 1982) (and citations therein).

21 *See generally* Orbital Locations, 54 P.&F. Radio Reg.2d 550 (1983); Orbital Spacing, 54 P.&F. Radio Reg.2d 577 (1983); Robinson, *Regulating International Airwaves: the 1979 WARC*, 21 VA. J. INT'L L. 1, 44 (1980).

22 *See, e.g.*, FMC v. Aktiebolaget Svenska Amerika Linien, 390 U.S. 238, 240–44 (1968); Network Project v. FCC, 513 F.2d 786 (D.C. Cir. 1975). Cf. City of Lafayette v. Louisiana Power & Light Co., 435 U.S. 389, 406 (1978).

carrier systems simply to foster competition.[23] It must instead make affirmative public interest findings that competition, for example, will spur technological progress, increase efficiency, and more rapidly expand customer choice.[24]

Executive Responsibilities

In addition to the limitations on entry and competition contained in titles II and III of the Communications Act, section 102(d) of the Satellite Act recognizes the foreign policy, trade, and national security aspects of international satellite communications and provides that the President is responsible to determine whether additional international satellite systems are required to meet unique governmental needs or are otherwise required in the national interest.[25]

The term »national interest« is not defined in the Satellite Act, but it encompasses considerations broader than those implicit in the FCC's regulatory »public interest« standard,[26] a standard which the courts have ruled is not limitless.[27] »National interest« is within the mandate of the Executive branch and includes such factors as general competition policy, whether entry will advance technological progress and innovation, promote U.S. international trade in goods and services, expand the international communications options available to the U.S. business community, and further overall U.S. spectrum management goals. Foreign policy and national security considerations are also important aspects of the national interest, and matters which are the Constitutional responsibilities of the Executive. The FCC in the past has generally deferred to Executive branch views on policies which are not directly within its regulatory purview.[28] In sum, the »national interest« standard in the 1962 Satellite Act should be read as according the Executive branch responsibility to determine the compatibility of proposed international satellite systems with the broad range of U.S. programs and policies affected by such enterprises.

International Obligations

In addition to the limitations on competition implicit in the 1934 Communications Act and the special »national interest« criterion in the 1962 Satellite Act, U.S. international obligations are relevant.

23 *See, e.g.*, FCC v. RCA Communications, 346 U.S. 86, 93 (1953); Hawaiian Teleph. Co. v. FCC, 498 F.2d 771, 778 (D.C. Cir. 1974).

24 *See* United States v. FCC, 652 F.2d 72, 91, 98–99 (D.C. Cir. 1980) (en banc). *See generally* Van Deerlin, *The Proposed Deregulation of Domestic Common Carrier Telecommunications*, 69 CAL. L. REV. 455 (1981); Palenberg, *International Telecommunications: Proposed Deregulation of Overseas Services*, 23 HARV. INT'L L.J. 214 (1981).

25 »Unique governmental needs« are not at issue here. None of the applicants now before the FCC maintains that its system will meet such needs nor has any agency identified unique needs that might thus be served.

26 *See, e.g.*, Domestic Satellites, 22 F.C.C.2d 86, 133 (App. D) (1970); Authorized Users, 6 F.C.C.2d 593, 594–95 (1962). *See generally* Legislation Note, *The Communications Statellite Act of 1962*, 76 HARV. L. REV. 388, 389 (1962). *Cf.* Telemanson v. United States, 386 F.2d 811, 812 (1st Cir. 1967); Gardels v. CIA, 484 F. Supp. 368, 371 (D.D.C. 1980).

27 *See* NAACP v. FPC, 425 U.S. 662, 669 (1976); National Organization for Women v. FCC, 555 F.2d 1002, 1017 (D.C. Cir. 1977).

28 *See, e.g.*, United States v. FCC, 652 F.2d 72, 90 (D.C. Cir. 1980) (en banc); AT&T Co. (NE Corridor Light Guide System), 51 P.&F. Radio Reg.2d 717, 725 (1982).

The INTELSAT Agreement entered into force for the United States on February 12, 1973.[29] While the INTELSAT Agreement implicitly acknowledges that nations party to the Agreement retain the sovereign right to establish satellite telecommunications facilities separate from the INTELSAT system, the Agreement establishes: (1) a generalized obligation of the parties to act in a manner consistent with and in furtherance of the principles stated in the Preamble and other provisions of the Agreement (Article XIV(a)); and (2) a consultation process to be undertaken before a nation or its designated operating entity (a »Signatory«) establishes, acquires, or utilizes separate, non-INTELSAT space segment facilities to meet its telecommunications requirements (Article XIV).

Article XIV(d) of the INTELSAT Agreement addresses the consultation obligation with regard to international public telecommunications services. In substance, it provides that a nation member or its Signatory shall furnish all relevant information to INTELSAT and shall consult with INTELSAT: (1) to ensure technical compatibility of the contemplated satellite facilities with the use of the radio frequency spectrum and the geostationary orbital space by the existing or planned INTELSAT satellites; and (2) to avoid significant economic harm to the global system of INTELSAT. At the conclusion of the consultation process, the INTELSAT Assembly of Parties (the principal organ of INTELSAT, composed of the representatives of all member nations) makes findings in the form of recommendations on the subjects of the consultation and further regarding the assurance that the proposed satellite facility will not prejudice the establishment of direct telecommunications links through the INTELSAT space segment among all the participants in the proposed system.

Considering the wide participation in INTELSAT, most of the contemplated separate satellite systems would involve two or more INTELSAT members. It is common practice for the INTELSAT members contemplating the establishment of separate satellite facilities to meet their international public telecommunication consultation requirements by consulting jointly with INTELSAT in accordance with the provisions of Article XIV(d) of the Agreement.

The term »public telecommunications services« is defined in Article I(k) of the INTELSAT Agreement as meaning:

> [F]ixed or mobile telecommunications services which can be provided by satellite and which are available for use by the public, such as telephony, telegraphy, telex, facsimile, data transmission, transmission of radio and television programs between approved earth stations having access to the INTELSAT space segment for further transmission to the public, and leased circuits for any of these purposes; but excluding those mobile services of a type not provided under the Interim Agreement and the Special Agreement prior to the opening for signature of this Agreement, which are provided through mobile stations operating directly to a satellite which is designated, in whole or in part, to provide services relating to the safety or flight control of aircraft or to aviation or maritime radio navigation.

23 UST 3813, 3816

At least one of the current U.S. applicants has contended that consultation with INTELSAT should not take place pursuant to Article XIV(d) but rather pursuant to

29 *See* Agreement Relating to the International Telecommunications Satellite Organization »INTELSAT«, 23 UST 3813, TIAS No. 7532 (1973). *See also* Simsarian, *Interim Arrangements for a Global Commercial Communications Satellite System*, 59 AM. J. INT'L L. 344 (1965).

Article XIV(e). A consultation pursuant to Article XIV(e), which deals with »specialized telecommunications services requirements,« would not include the subject of possible significant economic harm to the global INTELSAT system. A proposed satellite system, however, may well provide »public telecommunications services« (as defined in the INTELSAT Agreement) even though the applicant characterizes its endeavor as a noncommon carrier, and therefore, »private« satellite system.

Consultation within the INTELSAT framework occurs pursuant to the definitions contained in the INTELSAT Agreement. U.S. domestic communications policy currently recognizes a number of distinctions between traditional common carrier and other communications services.[30] Such distinctions, however, do not necessarily determine the international obligations of the U.S. Government. The United States will continue to consult with INTELSAT pursuant to Article XIV(d) regarding those satellite systems which would provide »international public telecommunications services.«

Consultation pursuant to the INTELSAT Agreement need not be protracted. Indeed, Article XIV(f) provides that INTELSAT shall make its recommendations within a period of six months from the date of commencing the consultation procedures. In practice, however, such a consultation cannot commence until the U.S. Government or the U.S. Signatory (Comsat) furnishes INTELSAT with all relevant information. In the past, delays in consultation have occurred because the information required of an applicant by the FCC in making its regulatory decision on initial authorization is not identical to that information relevant to the consultation with INTELSAT.

The United States is committed to ensuring that non-INTELSAT satellite systems are technically compatible with existing and planned INTELSAT satellites, and to avoiding significant economic harm to the global INTELSAT system. Accordingly, the Executive branch will initiate consultations with INTELSAT only for those non-INTELSAT systems which it believes meet the technical and economic conditions described in the INTELSAT Agreement. The United States will continue to consult with INTELSAT in good faith; therefore, the possibility cannot be excluded that, following the consultation process, the Executive branch might find that final regulatory authorization should not be granted.

The consultation with INTELSAT would normally end with a recommendation being made by the INTELSAT Assembly of Parties. Such recommendations are not binding on the United States, although the U.S. Government will carefully consider all recommendations. It will go forward only with systems it deems consistent with its obligations to INTELSAT.

III. *Foreign Policy Considerations Regarding International Satellites*

Addressing the issues raised by the proposed establishment of U.S. international satellite systems separate from INTELSAT requires consideration of U.S. foreign policy objectives. These objectives have been considered within the Executive branch and do not constitute an appropriate matter for independent determination by a regulatory

30 *See, e.g.*, National Assoc. of Regulatory Util. Comm'rs v. FCC, 525 F.2d 630, 640 (D.C. Cir. 1976); National Assoc. of Regulatory Util. Comm'rs v. FCC, 533 F.2d 601, 608 (D.C. Cir. 1976). *See also* CCIA v. FCC, 693 F.2d 198 (D.C. Cir. 1982).

agency. Here, however, the major foreign policy matters that were weighed are generally discussed to further understanding of the President's determination.

In his September 1983 letter to Chairman Charles H. Percy of the Senate Committee on Foreign Relations, Secretary of State George P. Shultz reiterated the basic foreign policy objectives of the United States in international communications, and they are similar to those enumerated in detail above: »To promote an environment in which ideas and information can flow freely among nations, to support the advancement of international commerce through the efficient and innovative use of communications resources, and to expand information access and communications capabilities of developing countries.«

The 1962 Satellite Act reflects these objectives and others which have been furthered through our participation in developing and supporting the INTELSAT system. INTELSAT's manifest success has:

- Provided a dramatic example of U.S. leadership in the peaceful use of space in the interest of all countries;
- Contributed to meeting evolving U.S. commercial needs for efficient international communications services;
- Provided developing countries with improved communications at reasonable and affordable rates;
- Confined the Soviet INTERSPUTNIK system to a relatively small portion of the world;
- Supplied developing countries with access to the geostationary, satellite radio frequencies; and,
- Provided benefits to U.S. companies through open international procurement for the international system's space communications equipment and services.

Permitting U.S. international satellite systems separate from INTELSAT, however, could:

- Bring new diversity and flexibility to international communications;
- Create or expand markets in new areas, such as customized, data, and video services;
- Provide incentives for INTELSAT and its Signatories to be more efficient and innovative; and,
- Permit outside financial sources to undertake high-risk, speculative ventures, thereby enabling INTELSAT to concentrate its resources on further extending basic services through prudent financial management.

To attain the optimal combination of benefits from the INTELSAT and additional U.S. international satellite systems, the United States must develop procedures and conditions under which procompetitive domestic goals can be made compatible with foreign policy objectives which have been well served by INTELSAT.

Background of INTELSAT

The United States played a leading role in the creation of INTELSAT in order to further national political, economic, and security objectives. The decision to speed development of communications satellites was first made by President Eisenhower and became a centerpiece of overall U.S. space and foreign policy programs. Five weeks after the Soviet Union launched the first man into orbit, President Kennedy sent his »man-to-the-moon« message to Congress (May 25, 1961). As part of an expanded U.S. space program, he called for accelerated development of satellites for worldwide communications.

Explaining to Congress the need for an international communications satellite system, Department of State officials in July 1961 emphasized:

- *The global concept.* The system should cover not only developed but also developing countries and service both small-volume as well as large-volume users, thus linking the United States to as many other nations as possible.
- *Political benefits.* The system should provide an opportunity for cooperation with as many other countries as possible in the peaceful use of space, thereby forging mutually beneficial ties.
- *Trade benefits.* The system should facilitate transaction of the world's business and ensure more open markets for U.S. technology and other products.
- *Conservation of the frequency spectrum.* The system should conserve, rather than consume, frequencies and thus help all nations, working through the International Telecommunication Union (ITU), make more effective use of the limited frequency spectrum.
- *National security.* Space communications should link U.S. forces and those of U.S. allies, and help in UN peacekeeping efforts.

These objectives were incorporated in the 1962 Satellite Act which declared in its Preamble that:

> [I]t is the policy of the United States to establish, in conjunction and in cooperation with other countries as expeditiously as practicable, a commercial communications satellite system, as part of an improved global communications network, which will be responsive to public needs and national objectives, which will serve the communications needs of the United States and other countries, and which will contribute to world peace and understanding. The new and expanded telecommunications services are to be made available as promptly as possible and are to be extended to provide global coverage at the earliest practicable date. In effectuating this program, care and attention will be directed toward providing such services to economically less developed countries and areas as well as those more highly developed, toward efficient and economical use of the electromagnetic frequency spectrum, and toward the reflection of the benefits of this new technology in both quality of services and charges for such services.

47 U.S.C. 701(a)–(b).

The concept of a global system was fundamental to meeting these policy goals. Technology transfer, aerospace product sales, and satellite launch policies evolved in keeping with the global system concept.

The United States has continued to reaffirm its strong commitment to INTELSAT over the years. Under Secretary of State James L. Buckley, after consulting with other parts of the Executive branch, wrote to the FCC on July 23, 1981, regarding transborder satellite services and reaffirmed the importance to the United States of the integrity of the INTELSAT system, stating:

> The foundation of our international communications satellite policy includes the concept of a global system to which all nations can have nondiscriminatory access, and through which international communications can flow free of artificial constraints.

As the same time, Mr. Buckley recognized that exceptional circumstances might warrant the use of domestic satellites for international service. The FCC was informed that there were no foreign policy objections to U.S. domestic space systems being allowed to provide transborder service to Canada, Mexico, or the Caribbean, provided there was consultation with INTELSAT under Article XIV and appropriate foreign government approval was obtained. Services could also be inaugurated if proposals are »supported by the U.S. Government and both the United States and the foreign governmental authorities concerned, in the absence of a favorable recommendation by the Assembly, consider in good faith that the obligations under Article XIV have been met.«

Service to Developing Countries

A primary foreign policy dimension of INTELSAT is service to developing countries. INTELSAT is a cooperative whose members make capital contributions commensurate with their use of the system. Members receive a return on capital (currently about 16 percent) and pay charges which reflect the variable costs of providing them service, together with an allocation of joint and common overhead costs. From the outset, INTELSAT has charged uniform rates for identical services provided on a global basis, although traffic in the Atlantic Ocean area reportedly is some six times that of the Pacific Ocean area and three times that of the Indian Ocean area.

There is volume efficiency in the use of communications satellites that has not been fully reflected in INTELSAT's rates. Such pricing policies further interests of the United States and other developed countries, as well as the interests of developing nations, because they promote the objective of linking as many countries as possible to the global system.

Although INTELSAT continues to introduce advanced equipment, it maintains less sophisticated technologies in service as well, to meet the needs of its less developed members. INTELSAT strikes a balance in all of the frequency ranges and provides for automatic and semi-automatic signaling and switching apparatus. How well INTELSAT can maintain and expand its ability to provide basic services, including the introduction of thin-route services such as VISTA and INTELNET, while, at the same time, attempting to meet all the demands of new specialized markets and services, is another consideration in examining the best ways to fulfill the intent of the INTELSAT Agreement.

Concerns were expressed by some administrations from developing countries at the April 1983 meeting of INTELSAT Signatories in Bangkok and again at the October 1983 meeting of the Assembly of Parties in Washington that if significant traffic were diverted from INTELSAT's Atlantic Ocean region to non-INTELSAT satellite systems, a worldwide rate increase might ensue. The avoidance of significant economic harm to the global system of INTELSAT by the conditions placed on non-INTELSAT U.S. satellite systems should allay those concerns.

Statements made by developing country representatives at the October 1983 meeting reflected their interpretation of the term »single global system« used in the INTELSAT Agreement. Some maintained this term precludes the establishment of virtually any satellite system outside INTELSAT and, indeed, would preclude even the existing »regional« satellite systems. The Preamble of the INTELSAT Agreement, however, itself envisioned »a single global ... system *as part of* an improved global telecommunications network« (emphasis supplied) and the Article XIV mechanism expressly contemplates non-INTELSAT satellites. Non-INTELSAT satellite systems today provide international public telecommunications services after appropriate consultation with INTELSAT. It has been suggested by some administrations that the development of additional satellite systems apart from INTELSAT on the part of the United States would contravene the INTELSAT Agreement and therefore constitute a signal that the United States no longer supports INTELSAT. This is clearly not the case.

Access to the Geostationary Orbit

How all nations can enjoy »equitable access« to the geostationary satellite orbit and to the associated radio spectrum is a major concern within the International Telecommunication Union (ITU). The results of the ITU's consideration of this issue at the upcoming World Administrative Radio Conference on the Use of the Geostationary Satellite Orbit and the Planning of the Space Services Utilizing It (Space WARC) in August 1985 and June 1988 is [sic] important to the United States and many other countries. INTELSAT's role in meeting developing countries' communications needs could make it a critical, if indirect, participant in the resolution of this issue on terms acceptable to ITU member nations.

For more than a decade, some developing countries have sought a guaranteed share of the geostationary orbit and the radio spectrum allocated to space services. They maintain that unconstrained growth of commercial satellite communications systems could exhaust the geostationary orbit and the frequencies currently available. Fearful of losing their share of what they understand to be limited global resources, developing countries in 1973 inscribed »equitable access« provisions into the ITU Convention. By the 1979 WARC, they were determined to write new rules for the use of the geostationary orbit and associated radio spectrum and obtained a commitment for the two-part Space WARC in the 1980s.

The availability of INTELSAT has not eliminated developing country demands for equitable access to the geostationary orbit and related spectrum. Nevertheless, its existence offers an alternative to the implementation of costly national satellite systems. So long as low-cost and technically attractive service is available through an international organization which accommodates the sovereignty interests of each country, there is added hope that developing countries may meet some of their needs through INTELSAT.

The proliferation of communications satellite systems already in progress, moreover, will heighten the importance of INTELSAT's role in frequency conservation. Increasing demand for the radio spectrum is hastening the development and implementation of innovative technologies which expand the capacity of the geostationary orbit resource and permit greater efficiency through multiple uses of the same frequency. Large-scale space platforms and other techniques have the potential to increase frequency usage efficiency by perhaps 50- to 100-fold; INTELSAT's multinational consolidation of demand – domestic, regional, and transoceanic – will thus have particular attraction. With these considerations in mind, the United States lent strong support at the October 1982 INTELSAT Assembly of Parties to the principle of domestic service using INTELSAT facilities, despite European opposition.

An indication of the developing countries' growing stake in INTELSAT can be found in INTELSAT's evolution toward playing a larger role in the provision of domestic satellite service. In 1974, Algeria proposed to lease INTELSAT capacity for enhancement of its domestic telecommunications network. Today, some 26 countries use INTELSAT to provide domestic service. INTELSAT has responded to this demand by committing itself to include [sic] planned domestic capacity, as opposed to relying solely on preemptible, spare capacity, in future generations of satellites. It has also developed higher power satellites that are compatible with the small earth stations that have proved most economical for domestic service.

INTERSPUTNIK

The Soviet Union uses satellite communications to help cement its relations with client states and to extend its influence with nonaligned nations. INTERSPUTNIK serves a number of Soviet policy goals, including Soviet interest in a »new world information order.« The success of INTELSAT in providing quality service at decreasing rates to developing countries has preempted the USSR from extending its technically inferior rival service, INTERSPUTNIK, to more than a few noncommunist nations. Since it began operations, INTERSPUTNIK has obtained only five new members (Vietnam, Afghanistan, Laos, South Yemen, and Syria) beyond its original nine charter members (Bulgaria, Cuba, East Germany, Poland, Romania, Mongolia, Czechoslovakia, Hungary, and the USSR). Other countries using the INTERSPUTNIK system include Algeria, Libya, and Nicaragua.

Some suggest that potential competition with INTELSAT will lend impetus to the development of INTERSPUTNIK and increase Soviet influence in international satellite communications.[31] Soviet development of INTERSPUTNIK, as well as the emergence of the Soviet Union as a competitor in the world market for satellite launch services, however, is likely to go forward unaffected by U.S. decisions to authorize additional U.S. international satellite systems. A substantial weakening of INTELSAT as the dominant global satellite communications system, nevertheless, could potentially enhance Soviet efforts to penetrate developing countries through Soviet communication satellite facilities. An INTELSAT less attentive to developing country needs could encourage INTERSPUTNIK's efforts to expand its service area. This will continue to be an area of concern under any circumstances.

Satellite Proliferation

In addition to INTERSPUTNIK, other systems outside INTELSAT have evolved, as earlier indicated. In 1978, for example, governments (including the United States and the Soviet Union) founded the International Maritime Satellite Organization (INMARSAT) to provide service to ships at sea. A number of regional and domestic satellite systems have also developed, after consultation took place with INTELSAT under Article XIV(d) of the INTELSAT Agreement.

In the future, INTELSAT is likely to face additional satellite competition. There are a number of existing and planned satellite systems in addition to any U.S.-based systems which may be authorized. These include:

31 *See, e.g.*, Statement of Mr. Joseph Charyk, Chairman, Comsat Corporation, before the Senate Foreign Relations Committee, 98th Cong., 1st Sess. (Oct. 31, 1983) at p. 8.

– *UNISAT.* Britain's first satellite system could provide both domestic television transmission and certain international communications services. After its scheduled launch in 1986, UNISAT's beam or »footprint« will cover the U.S. eastern seaboard as well as most of Western Europe.
– *TELECOM.* The first TELECOM satellite, the French counterpart to UNISAT, was launched in 1984. The system not only will serve domestic French needs but also will cover most of Europe and provide telephone and television connections to the French overseas departments. Its capacity will extend to the French Caribbean, eastern Canada (St. Pierre and Miquelon), and the Indian Ocean (Reunion and Mayotte).
– *EUTELSAT.* This Paris-based consortium of 20 participating European countries launched its first communications satellite in 1983. The system will provide telephone, television program distribution, and data transmission services within Western Europe, North Africa, and the Middle East countries bordering on the Mediterranean Basin.
– *Arabsat.* The Arab Satellite Communications Organization, based in Riyadh, will serve 22 Arab countries. The first of its two satellites will be launched in 1985. The system is designed to supply telephone, television distribution, and data transmission services to most of the Middle East and North Africa.
– *Palapa.* Indonesia's Palapa System currently provides both domestic service as well as service with the Philippines, Malaysia, and other Southeast Asia nations.
– *Pacific Basin Proposal.* At the Pacific Telecommunication Conference in January 1983, the Japanese Research Institute of Telecommunications and Economics (RITE) presented a detailed plan for a Pacific Regional Satellite Communications System, ostensibly designed to »supplement« the existing INTELSAT network. It would provide two dissimilar services: high-speed digital communications for data and video transmission between major cities from the U.S. west coast to Japan, Australia, and Southeast Asia; and low-volume telephone communication between rural areas, remote islands, and their capital cities.

U.S. Role in INTELSAT

The U.S. role in INTELSAT continues to be strong, although it has changed over the past 20 years. The U.S. investment share has decreased from 61 to 23 percent; hence the U.S. weighted vote in the Board of Governors has decreased to the current 23 percent. An international secretariat of some 600 INTELSAT staff now manages the system rather than Comsat. A U.S. citizen was recently elected Director General of INTELSAT. INTELSAT not longer purchases almost all of its equipment from U.S. manufacturers, although the United States still supplies about 70 percent of INTELSAT's purchases. The United States is the host country for the INTELSAT headquarters.

The United States has been and should continue to be a strong leader and contributor to the INTELSAT system. Changing technology, competitive economics, and diversifying user needs, however, have created a new international telecommunications environment. There is a manifest trend toward coexistent, separate national and regional satellite systems. This does not obviate the continued need for a global system providing an essential core for public-switched international communications. The 1962 Satellite Act and the INTELSAT Agreement both specifically anticipated communications satellite systems outside INTELSAT, and provided the flexibility to allow for and to respond to such systems.

INTELSAT serves the world well. It has established and currently operates an efficient global communications system; promotes closer ties among noncommunist countries; facilitates international business expansion; helps to develop markets for U.S. industry; prevents the spread of a global communications satellite network controlled by the Soviet Union; and is an effective international organization reflecting shared technical and political interests. At the same time, new satellite systems can supply services inconceivable 20 years ago and provide services sought by high-volume users, including the U.S. Government. New approaches promise diversity and flexibility.

INTELSAT faces growing competition from new fiber optic cables, which may constitute a more significant challenge to it than separate satellite systems. The transatlantic cable (TAT-8) planned for 1988 by AT&T and the IRCs, Teleglobe Canada, and European telecommunications administrations will have a capacity equivalent to about 38,000 telephone circuits, as previously noted, and nearly quadruple the current submarine cable capacity across the North Atlantic. This fiber optic cable, moreover, will have technical capabilities, including the ability to transmit high-quality video signals, which existing submarine cables lack.

A »status quo approach« often has short-term appeal and merit from a foreign policy standpoint. Change inherently creates pockets of concern in the complex environment of international relations. By its very nature, however, telecommunications is uniquely amenable to change. The issues associated with international telecommunications cannot and will not stand still. They are driven by technology – and technology, in turn, is driven by continuing innovation and evolution.

U.S. policy leaders 20 years ago could not easily have envisioned the exponential expansion of communications horizons through new technology which has subsequently occurred. They did, however, anticipate the need for flexibility to develop the then-uncharted telecommunications frontier.

Unlimited proliferation of communications satellite systems separate from INTELSAT has the obvious potential to inflict significant economic harm on the global system. At the same time, U.S. economic goals require recognition of the changing marketplace and encouragement of innovation. The approach discussed in this report and reflected in the President's determination strikes a sound balance in this regard.

INTELSAT as a Competitor

An essential ingredient for the formation of INTELSAT was the provision on universal pricing for each defined service that is contained in Article V(d) of the INTELSAT Agreement. The Board of Governors, under the guidance of the Meeting of Signatories, establishes rates for each specific service or group of services which are then applied on a nondiscriminatory basis. The Signatories have established the principle that rates shall, as far as practicable, reflect costs. This built-in flexibility within the INTELSAT Agreement permits INTELSAT to offer new services, to take advantage of new technologies, and to price new services as close as practicable to cost (including direct as well as indirect costs).

This flexibility lessens some of the concerns which arose domestically when long-distance competition was sanctioned, but incumbent carriers were not afforded the ability to price responsively.[32] INTELSAT's ability to match the prices of other international satellite systems, however, is limited as it deals through its Signatories. As indicated above, INTELSAT's charges constitute only part of the end-user price for service.[33] Significant changes in end-user prices are thus dependent on action by its Signatories (or, in the United States, by Comsat and terrestrial carriers such as AT&T).

INTELSAT, in any event, should enjoy some competitive advantages with respect to new satellite systems offering customized services. The INTELSAT system may embody economies of scale and scope; INTELSAT enjoys a breadth of coverage today that new satellite systems could not easily replicate. The technology currently used by INTELSAT may not permit the organization to provide efficiently all of the customized services some of the new entrants envision. INTELSAT may also, as a matter of prudent management, choose not to seek to offer all such services. New satellite entry subject to the conditions discussed in this report, however, does not pose any substantial risk of significant economic harm to the INTELSAT global system.

IV. *Recommended Approach to New System Proposals*

The primary focus of this report is on those factors underpinning the President's November 1984 determination that new U.S. entry into the international satellite business is »required in the national interest,« provided entrants are not interconnected with public-switched message networks and joint consultation with INTELSAT is undertaken. The Presidential determination does not constitute endorsement of any specific pending satellite application. It represents, rather, a determination of the terms and conditions under which entry will be in the national interest. Reducing barriers to entry and permitting entrepreneurs to go forward is an important step toward achieving an efficient market for customized services. Other policy components to this process would also facilitate efficiency and can be pursued in parallel proceedings. In this section, the Presidential determination and those »parallel track« matters are discussed in detail.

32 *See* Aeronautical Radio, Inc. v. FCC, 642 F.2d 1221, 1228–29 (D.C. Cir. 1980); *cf.* National Assoc. of Greeting Card Publishers v. U.S. Postal Service, 569 F.2d 570, 582 (D.C. Cir. 1976) *rev'd in part*, 434 U.S. 884 (1977).

33 Some indication of the costs associated with current arrangements is afforded by considering the minimum cost of a one-hour video transmission from Rockefeller Center, New York, to the British Broadcasting Corporation facilities in London. At present, the minimum cost for such service would be $2,727 per hour. On the U.S. side, AT&T's charges for domestic transmission (New York to Andover, Maine) would be about $439. Comsat's minimum charge would be about $633 (of which $480 goes to INTELSAT), yielding a total U.S. cost of about $1,072 per hour. British Telecom would then charge 4,200 gold francs for the British side of the circuit (which charge would include landline charges) of about $1,655 per hour (of which INTELSAT would get another $480). Thus, of this total charge of $2,727, INTELSAT would receive $960, or about 35 percent. The figures, it should be noted, do not necessarily reflect INTELSAT's payments to owners. Satellite charges were computed using Comsat's »Satellite Television International Tariff Information Handbook« (Aug. 1, 1982, as revised) at pp. 140, 142. Domestic AT&T tariff prices were supplied by AT&T.

New Systems Should Be Permitted

First, additional U.S. international satellite systems should be permitted, but subject to the terms and conditions previously specified. New service alternatives are proposed in the pending applications that would be in the national interest. These include certain international video and data transmission services not now available through the INTELSAT system. The proposed systems also may offer major users a means of enjoying more of the savings associated with service on high-traffic volume communications routes than those customers have today.

Users, and particularly sophisticated business service customers, stand to benefit from satellite communications options which are more closely tailored to their special needs. INTELSAT has concentrated on its primary function – serving public-switched service users. The present INTELSAT system, moreover, is not configured to provide every important customized business service efficiently. Requiring business users with special needs to conform to »lowest common denominator« communications offerings imposes economic costs which can and should be lessened.

Service Limitations Required

Service limitations are required, however, to avoid significant economic harm to INTELSAT. New entrants thus should be limited to the provision of customized service. Such services involve the sale or long-term lease of transponders or space segment capacity for communications that are not interconnected with public-switched message networks. Customized services include intracorporate networks and television transmission. Emergency restoration services would also constitute a customized service.[34] Prospective new satellite entrants maintain they will target communications needs that are not now efficiently served by INTELSAT. They should thus be authorized under regulatory terms and conditions that will hold them to their commitments and ensure that their attention is focused on serving and developing the customized service market.

At present, public-switched message traffic comprises the overwhelming majority of INTELSAT traffic. As indicated above, the most recently published INTELSAT annual report states that full-time voice, record, and data service accounts for about 86 percent of the total satellite utilization revenue INTELSAT receives. Such public-switched traffic constitutes the commercial core of the INTELSAT operation and, again as indicated above, it is forecast to increase by 15 percent over the 1988–2000 time period.[35] Technical advances including IDDD, as well as additional entry into the international telephone business by U.S. carriers such as MCI and GTE Sprint, should have a positive effect on public-switched traffic. Increasing service and price competition among AT&T and other U.S. carriers, moreover, are likely to stimulate overall demand. There is evidence suggesting such competition in domestic public-switched

34 Recommendation D.1 of the International Consultative Telegraph and Telephone Committee (CCITT) places certain limitations on customer use of international private leased circuits. FCC regulations do not now permit the resale or sharing of international private line services.

35 1983 INTELSAT Ann. Rep. at p. 17; 1985 U.S. INDUSTRIAL OUTLOOK at p. 31–37. *See* note 11, *supra*.

service markets stimulated demand.[36] There are also indications that this demand-stimulation effect may already be operating in some international public-switched service markets.[37]

Limiting new entrants to customized services reduces any likelihood of significant adverse economic impact on INTELSAT. Such restrictions are sustainable domestically and internationally, particularly given the multinational character of international telecommunications and the fact that foreign PTTs police the services provided by companies serving their countries. No regulatory regime can be »air-tight.« But the limitations discussed here are adequate to safeguard the economic integrity of INTELSAT, especially given public-switched market trends as discussed in the subsequent section on the sustainability of such restrictions.

Cost-based Access an Important Issue

The economic well-being of INTELSAT may be furthered by »cost-based« access for customized services. One way this could be secured is by permitting U.S. carriers and users to deal directly with INTELSAT, with the U.S. Signatory (Comsat) serving as their ministerial agent. Another way could be to ensure that all of the costs which Comsat and the carriers assess in addition to the basic INTELSAT charge reflect legitimate, necessary costs.

The process by which customers obtain international satellite communications service results in end-user prices substantially above INTELSAT circuit charges. The current U.S. arrangement where Comsat, in effect, functions as exclusive U.S. marketing agent for INTELSAT circuits, may be ill-suited to an era of proliferating customer demands. No single entity, no matter how perceptive, can reasonably be expected to anticipate and satisfy all customer demands and needs in a market which is experiencing rapid

36 Between 1978 and 1979, for example, U.S. domestic telephone revenues increased by 6 percent. In 1978, the remaining restrictions on competition among domestic carriers were removed. Between 1979 and 1980, the first full year of generally unrestricted public-switched message competiton, the annual rate of increase rose a full percentage point, to about 7 percent (using constant 1972 dollars). Between 1980 and 1981, the annual rate of increase rose to about 10.5 percent, or about 40 percent higher than the rate which prevailed when the domestic public-switched services market was far less competitive. *See* 1978 through 1982 U.S. INDUSTRIAL OUTLOOKS.

37 As one international communications expert has stated:

> As you may know, we've had some competition on service to Canada, and that same competitor [MCI] has set up an experiment with Australia. It's very early to be drawing direct conclusions, but I'd like to share some figures with you. We had forecast ten percent growth this year in our messages to Canada. Our actual growth in the first seven months of the year was sixteen percent – six percentage points higher than we predicted. Now, these results are subject to interpretation. They can be attributed to such things as marketing efforts, advertising, and, of course, the recovery of our economy. And, as I said, it is still very early. But it certainly appears to me that, from what we've seen so far at least, competition has not hurt growth. And perhaps, as more time elapses, we will be able to say it has stimulated business. It makes one wonder, if we had competition in other countries, whether perhaps the total communications package would grow at an accelerated rate.

Remarks of Mr. R.B. Nichols, AT&T, at TELEVENT '83, Montreux, Switzerland, Oct. 25, 1983 at p. 17.

demand-inducing and cost-reducing technological advances. Permitting expanded, direct, cost-based access to INTELSAT may be the most reliable means of substantially reducing costs and ensuring valid entry signals.

Recently, the FCC required Comsat to unbundle its INTELSAT tariff into separate, »cost based« rates for space segment and earth segment services.[38] In addition, the FCC determined that AT&T and the IRCs could own earth stations independent of the traditional joint ownership arrangement, subject to FCC approval on a case-by-case basis. This decision seeks to stimulate competition to provide earth station services, and to lower costs and increase the availability of services to the consumer. The decision may also allow the FCC further to identify legitimate cost components of Comsat's space segment rate.

The Executive branch shares the FCC's goals of providing users with cost-based international satellite communications services of high quality and reliability, tailored to individual needs. The FCC recently declined to commence a formal rulemaking with a view toward sanctioning expanded, cost-based access to INTELSAT. It expressed the view that regulatory measures could lessen the need for such structural change for end-users and Comsat's carrier customers, but emphasized that it was not foreclosing reconsideration of direct access should alternative measures prove ineffective.[39]

The Executive branch nevertheless recommends that the FCC examine cost-based carrier and user access to INTELSAT with respect to customized services, and the Department of Commerce will soon file detailed recommendations in this regard. While this issue might entail substantial public benefits when viewed in parallel with the establishment of alternative satellite systems, it is not a prerequisite for, nor should it be the basis for any delay in, ruling in the applications now before the FCC.

In sum, the President has determined that entry by additional international satellite systems, limited to customized services, is required in the national interest because it will:

- Provide users more flexible options and facilitate more efficient international satellite communications services;
- Promote development and use of satellite technology; and,
- Afford U.S. entrepreneurs an opportunity to develop new communications services and increase international trade opportunities.

V. *An Evolutionary Approach*

The concept of additional entry into the international satellite communications business is not new. The United States, as earlier discussed, has permitted such entry by sanctioning transborder satellite communications, after consultation with INTELSAT, and has supported establishment of a number of regional satellite systems. The approach recommended here should thus be regarded as facilitating evolutionary, not revolutionary, change in international telecommunications.

38 Earth Station Ownership (CC Dkt. 82-540), FCC Mimeo 84-605 (released Dec. 18, 1984).

39 Regulatory Policies Concerning Direct Access to INTELSAT (CC Dkt 82-548), 97 F.C.C.2d 296 (1984).

Objections have been voiced to any changes in the status quo. Some suggest, for example, that there would be a severe adverse economic impact on INTELSAT from new systems, even if the scope of their offerings were limited, or that any limitations would prove unenforceable or ineffective over time. Similarly, it has been suggested there are certain international radio frequency management obstacles. Finally, it has been asserted that U.S. international trade or other interests, or the legitimate interests of less developed countries, could be adversely affected. None of these objections withstand close analysis, however, nor do they override the advantages of additional entry to the national interest.

No Adverse Economic Effects Are Likely

Under the recommendations and criteria discussed in this report and in the President's determination, new satellite entrants could not offer public-switched services directly or indirectly and would be obliged to focus on developing customized service markets. Since public-switched services comprise by far the largest part of international traffic, any significant adverse impact on INTELSAT could result only if: (i) customized communications quickly supplant conventional services as the mainstay of the international communications business; (ii) such new services constitute a uniquely profitable line of commerce, the profits from which are essential to subsidize other necessary but unprofitable INTELSAT undertakings; and (iii) INTELSAT proves unable effectively to match new entrants, by, among other things, achieving end-user price reductions, broadening its service repertoire, and providing carriers and users direct access options. Virtually all of the Executive branch's analysis, however, indicates that these possibilities are remote.
According to INTELSAT forecasts (see Table I), in 1988 traffic on its transatlantic voice-grade circuits will continue to be composed overwhelmingly of message telephone service (MTS) and related public-switched services. Specifically, of 15,603 satellite voice-grade circuits to 18 major European countries planned in 1988, INTELSAT has forecast 14,000 will be used for MTS alone. Under the Executive branch approach, new entrants would thus be barred from providing services which are directly competitive with some 90 percent of INTELSAT's voice-grade offerings, according to INTELSAT's own estimates.

Table I 1988 INTELSAT Traffic Projections

INTELSAT has forecast the following breakdown on its 1988 voice-grade traffic to 18 major European countries:

14,185	MTS
113	Record Service
1,259	Alternate Voice Data (AVD)
46	Data
15,603	Total voice-grade (4 kHz) circuits

INTELSAT has also projected for 1988 the following numbers of channels for its International Business Service (IBS):

15	1.544 megabit per second (MBS) channels-K band
182	56/64 kilobit per second (KBS) channels-K band

For 1988, INTELSAT forecasts seven television transponder leases to Europe.
Source: INTELSAT *Global Forecast* (June 1982).

The impact of new entry on markets for other than public-switched services will depend on growth in demand for those customized services and users' evaluation of the relative merits of the rate and service options offered by the entrants and the incumbent, INTELSAT. Because of the dynamics of the international communications marketplace, uncertainties regarding user needs and preferences, and imperfect knowledge of the likely pricing strategies of entrants and INTELSAT alike, any forecast of market capture by the new entrants and possible revenue loss by INTELSAT is subject to risk of wide error. Review of several market penetration and growth scenarios, however, indicates that substantial economic harm to INTELSAT from new entrants limited to private non-switched services is highly improbable. Any traffic diversion and loss of business revenue from INTELSAT to the entrants will almost certainly prove less than the expected growth in revenues from users of INTELSAT services. The total annual revenues most likely to be obtained by the proposed entrants, moreover, will not have significant adverse effects on INTELSAT or its rates for switched services.[40]

Some further contend that new satellite system entry will result in widespread and substantial de-averaging of INTELSAT's prices, with the further consequence that »thin-route« prices will rise abruptly while »thick-route« prices rapidly fall. This, critics maintain, will result in sharp increases in communications costs for developing countries who today are said to benefit from internal, INTELSAT-devised and -administered cross-subsidization schemes. Such pessimistic forecasts, of course, are comparable to those which were advanced when U.S. domestic competitive new entry was under consideration by the FCC.

Possible adverse effects on developing nations are of significant concern, given the increasingly important role communications plays as a catalyst for overall economic development and given the United States' longstanding commitment to improving the economic prospects of developing nations. Analysis indicates there is little possibility of significant adverse effects on INTELSAT, or, in turn, adverse effects on developing nations.

There are three reasons for this conclusion. First, by far a majority of INTELSAT's core revenues and its basic service functions would be »off-limits« to new entrants. Second, even assuming some significant cross-elasticity or interchangeability of demand between customized and conventional services, both markets currently are growing rapidly. Revenue »siphoning« is likely to occur, if at all, only when the markets at issue are static, which is not true here. INTELSAT's charges, moreover, typically constitute but part of end-user charges for communications circuits. Increases in INTELSAT's

40 *See generally* Technical, Economic, and Institutional Feasibility of Customer Premises Earth Stations for INTELSAT Services, (NTIA: M/A-Com. DCC, Inc., May 1983); Present and Projected Business Utilization of International Telecommunications (NTIA, 1981).

charges for public-switched offerings, which are unlikely, need not necessarily be reflected in higher end-user circuit prices. Third, INTELSAT is in a good position to compete. The organization has an extensive array of advanced spacecraft, a highly talented technical and managerial cadre, and enjoys global acceptance and presence. These are potential competitive advantages few entrants could hope to replicate.

In sum, while potential adverse effects of new entry on developing nations' communications prices is an issue, there are few foreseeable conditions, if any, under which the pessimistic forecasts advanced in opposition to new entry might conceivably materialize. In the unlikely event such problems develop, moreover, there are a number of corrective measures available other than pursuing unnecessary restrictive entry policies.

No Valid Spectrum Management Objections

Concern has been advanced that U.S. approval of additional international satellite systems could complicate international radio spectrum management programs. Such approval allegedly could be perceived as inimical to the goal of ensuring »equitable« access to and use of the geostationary orbit and associated radio spectrum, increasingly regarded as a scarce and valuable international resource, and thus compromise our efforts to ensure international acceptance of flexible orbit and spectrum regulation. These are not unreasonable concerns for study. Our review of the possible effects of such U.S. action, however, suggests little adverse impact on radio frequency management policies and programs.

The U.S. international satellite systems now being considered by the FCC propose to use current technology and to function in the frequency bands allocated internationally for such services. The proposed uses accord with applicable international radio regulations, as do the projected power flux density, »station keeping,« and »pointing accuracy« features of the proposals. Engineering review of the proposed new systems indicates they would comply with pertinent international radio regulations.

Questions have been raised regarding the possible effect of U.S. approval of additional satellite systems on current and future international discussions of geostationary orbit use, previously discussed in the part of this report surveying foreign policy concerns. The orbital positions proposed by the new entrants will require technical coordination under the ITU Radio Regulations, and the systems must eventually be recorded by the International Frequency Registration Board. A preliminary review indicates all of the proposed positions can be accommodated through the current ITU process.

Since the advent of commercial satellite communications, there has been disagreement internationally between those favoring a flexible international regulatory approach, and those urging rigid, »a priori planning« of orbital resource use. The United States and other nations have favored a flexible approach to facilitate the evolution of satellite communications technology. Some foreign administrations, however, have pressed for a more rigid approach on the ground it will ensure »equitable access« to the geostationary orbit, especially on the part of developing countries.[41]

International discussions regarding orbital »slot« utilization antedate current proposals to deploy additional U.S. international satellite systems. These discussions will be an

41 *See* Robinson, *Regulating International Airwaves: The 1979 WARC*, 21 VA. J. INT'L L. 1, 44 (1981).

important part of future Space WARCs regardless of the disposition of the pending U.S. satellite system applications. Granting these applications could provide those favoring a rigid approach some additional support for their views; they may contend that the United States is using more than its »fair share« of what is perceived to be a scarce international resource. Such arguments, however, are not compelling.

The orbital positions sought by applicants for new U.S. satellite systems are unlikely to interfere with the rights of other nations to make use of orbital resources. Additionally, experience gained through such new systems would be available to other administrations and thus afford them a means of better serving their own national communications needs. Several of the U.S. satellite system applicants, moreover, propose the sale or long-term lease of space segment capacity which could afford both U.S. and foreign users an opportunity to invest directly in, and secure the benefits of, advanced satellite communications. Under the Executive branch approach, both U.S. and foreign customers would be offered new, potentially valuable, service options.

The United States, by taking a flexible approach toward orbit use, has managed to foster the development of new communications techniques which, in turn, have made possible steadily more intensive use of the orbital arc. Spacing between U.S. domestic satellites has been steadily reduced from 5 degrees to 2 degrees over the past decade, and advances in technology should aid in achieving even more intensive use. Such gains in technical sophistication and effectiveness would not have been accomplished as readily, if at all, had the United States adopted the rigid approach some nations urge.

These concerns on the part of some nations are being noted by policymakers in preparing for the 1985–88 Space WARC and international radio conferences generally. The restrained approach toward additional international satellite systems reflected in this report and in the President's determination, however, should ameliorate international concerns. It is possible to accommodate the interests of INTELSAT, new entrants, and, more importantly, the users of international communications both here and abroad, and thus to maximize the benefits afforded by space satellite technology.

Positive International Trade Effects

Related contentions have been advanced concerning U.S. approval of additional international communications satellite systems. It has been contended, first, that U.S. approval will dissipate U.S. influence over INTELSAT and, second, diminish the significance of INTELSAT as a major purchaser of U.S. aerospace products. Third, it has been contended that U.S. action will trigger a further proliferation of regional and transoceanic satellite systems sponsored by other nations which will rely chiefly on indigenous aerospace firms, thus gradually eroding any technological and commercial edge the United States enjoys in the aerospace field. Finally, some maintain that communications administrations abroad will seek to influence procurement decisions made by new U.S. satellite system entrants.

INTELSAT scheduled 12 INTELSAT V and V-A satellite launchings between 1982 and 1985. The total number of satellites in the current expansion program is 15, with an estimated value of $1.3 billion (including launch costs). In March 1982, INTELSAT awarded Hughes Aircraft Company a $700 million contract for the purchase of the first five satellites of the next generation, INTELSAT VI. Each INTELSAT VI satellite will have the capability to handle more than 30,000 telephone circuits and several television programs – more than twice the capacity of the latest INTELSAT V-A satellite – and a ten-year design life. INTELSAT will launch the first satellite in this series in 1986 aboard the U.S. Space Transportation System (Shuttle) and may use the European Space Agency's Ariane system for others. INTELSAT estimates the cost of this latest development program will reach $2.2 billion by 1992. U.S. aerospace firms anticipate participating in this program, and the Executive branch has no ground to assume this will not be the case.

It is not U.S. influence that currently affords U.S. aerospace producers a significant share of INTELSAT's procurement. The success U.S. producers enjoy is due chiefly to the superior quality of their products, the attractiveness of their prices, and the sophistication of their technology in what is increasingly a fiercely competitive world market. The INTELSAT Definitive Agreement, moreover, specifically mandates open and competitive procurement. It is unfair to imply the skilled professionals who comprise the INTELSAT Executive Organ would disregard the requirements for competitive bidding contained in the Agreement, overlook products offered by U.S. firms at competitive prices, and thus compromise a well-earned reputation for fair and business-like conduct of this important international enterprise.

At present, U.S. aerospace producers confront intensifying international competition from a diversity of high-caliber, multinational firms, and this trend is likely to continue independent of the decisions at issue here. U.S. firms enjoy some advantages in producing certain classes of spacecraft – large capacity spin-stabilized satellites, for instance. While U.S. firms are preeminent in the international aerospace field, fewer and fewer free world aerospace projects rely exclusively on products supplied only by one nation's firms. Extensive joint venture and cross-licensing arrangements are increasingly characteristic of this field.

INTELSAT has purchased from a broad range of suppliers, and the percentage of its procurement awarded U.S. firms has declined as the commercial competence of non-U.S. firms has grown. The United States, however, should not fear this increased competition. In a free trade environment, such competition provides a necessary and highly desirable spur to greater efficiency, more rapid innovation, and improved customer responsiveness. Indeed, much of the rapid growth in the U.S. aerospace business is attributable to the competitiveness of this field generally and the resulting incentives to perform efficiently.

The intrinsic talents and abilities of U.S. aerospace firms should not be adversely affected by U.S. approval of additional international satellite systems. INTELSAT's professionals will continue to abide by the competitive procurement requirements contained in the Definitive Agreement. The importance of INTELSAT as a purchaser of aerospace products, both of U.S. and foreign manufacture, in sum, should not be impaired.

The Executive branch has also seriously considered contentions that U.S. approval of new satellite systems could have a »domino effect« and trigger additional entry by »state-subsidized« European and other systems that will adversely affect INTELSAT and not make use of U.S. aerospace products. While the details of all such additional satellite systems are not yet available, a number of satellite systems are now functioning or planned worldwide in addition to U.S. systems as earlier discussed. Extensive submarine cable facilities, moreover, are also under construction.

Available information does not indicate U.S. aerospace firms have been foreclosed from competing to supply existing and planned regional satellite systems. Ford Aerospace, for example, reportedly is a major subcontractor for both Arabsat and French satellite systems. Hughes Aircraft has supplied spacecraft for the Indonesian

Palapa regional system and has longstanding relations with SPAR, the Canadian firm which is the prime contractor for the Brazilian domestic satellite system. Ford, Hughes, and RCA all have commercial arrangements with Japanese aerospace companies and thus stand to participate in any satellite systems which Japanese firms may propose in the Pacific region.

It is unsound to assume, moreover, that any sanctioning of new U.S.-based satellite systems will adversely affect INTELSAT since foreign entrants may not be subject to limitations such as those recommended for U.S. entrants. Having placed restrictions on the activities of U.S. entrants, the national interest would require comparable limitations on the services any foreign satellite system might provide to and from the United States.[42]

At present, the United States accounts for a majority of international telecommunications traffic and, indeed, is said to constitute some 40 percent of the total world communications services and products market.[43] Access to the U.S. market is thus commercially critical. Having taken appropriate steps to safeguard the economic integrity of INTELSAT, the United States would not sanction actions by foreign systems serving U.S. markets that would undermine our limitations and place U.S. firms at a competitive disadvantage.

Finally, the Executive branch has weighed the possibility some foreign governments might consider dictating procurement requirements in exchange for permitting non-INTELSAT satellite systems to access their markets. The United States would oppose any initiative by foreign administrations which would discriminate against U.S. aerospace firms. The United States does not wish to regulate the procurement decisions of noncommon carrier, satellite systems. The United States, nevertheless, would consider declining to consult on proposals involving unacceptable procurement provisions that could adversely affect competition in the aerospace industry. Such provisions might also raise questions under international trade agreements.

Not only is there little ground for concern that U.S. approval of limited entry into the international satellite field would adversely affect international trade, but there are also sound reasons to forecast positive consequences. International services today are priced considerably above domestic circuits of comparable length. At present, for example, MCI charges a minimum of $3,700 per month for a full-time, voice-grade private line between New York and London. A New York to Los Angeles private line circuit retails for from $1,507 (MCI), to $1,701 (Western Union) to $1,150 (RCA). International service, in short, costs between two and three times comparable U.S. domestic service. A U.S. firm offering international circuits at prices comparable to U.S. domestic prices should thus experience significant demand.

U.S. financial services and data processing companies constitute major factors in the international communications market, with annual communications bills amounting to tens of millions of dollars in several instances. Reductions in these communications costs

42 Under the 1921 Cable Landing Act (47 U.S.C. 34, 35) and the delegation of Presidential authority to the FCC in Executive Order 10530 (3 CFR 189 (1951–1958 Comp.)), the FCC enjoys broad authority regarding the provision of international services by foreign entities directly or indirectly to the United States and has authority to take steps to ensure equality of opportunity among U.S. and non-U.S. carriers in the international telecommunications business. Section 308(c) of the Communications Act. (47 U.S.C. 308(c)) empowers the FCC to place comparable requirements on those providing international services by radio.

43 *See* 1985 U.S. INDUSTRIAL OUTLOOK at p. 31–33.

imply lower business – and, ultimately, customer – costs and an expansion in business activity. New entrants may also offer large users services more closely tailored to particular corporate needs. Worldwide credit card and electronic funds transfer operations, for example, may be heavily dependent on the availability of efficient, dedicated satellite communications networks. New communications service options and resulting efficiency gains should be reflected ultimately in lower costs to consumers and, in the case of U.S. firms, enhance the attractiveness of their products in international markets.

New communications satellite offerings should also have an affirmative effect on the U.S. services sector generally, which is of special importance given the contribution this sector makes to U.S. overall foreign trade. In recent years, the services sector has become a major source of export receipts in U.S. balance-of-payments accounts. Included in this diverse sector are enterprises including data processing, engineering, architectural, and construction services, advertising services, management consulting and accounting services, insurance services, and the provision of video programs, all of which are increasingly dependent on the availability of effective and efficient international communications. The market for U.S. programs is particularly important given the rapid development of cable television, commercial television, and other video services in Europe. In 1982, receipts from services exports were $40.4 billion, about one-fifth the amount of U.S. merchandise exports. Over the past decade, growth in U.S. services exports has partially offset losses in merchandise export accounts. Services constitutes a key component of U.S. international trade, and expanding U.S. communications options should contribute to its growth.[44]

National Defense and Security Implications

International communications constitutes a critically important component of U.S. and allied defense and security programs. The U.S. Department of Defense is the largest single user of international communications services, spending more than $50 million annually for more than 220 commercial satellite channels. Moreover, the Defense Department not only has extensive North Atlantic Basin communications requirements; it also needs to communicate globally to remote locales and has relied significantly on the INTELSAT system in this regard. The Defense Department is concerned, therefore, that additional competition in the international satellite communications business not impair the cost-effectiveness or service quality of the INTELSAT system. Approval of additional U.S. international satellite systems, subject to the limitations discussed in this report, will not adversely affect national defense.

A key interest of the Defense Department and the national security community is ensuring the effectiveness and survivability of international communications services through redundant routing and maintaining a broad mixture of international communications facilities.[45] The Defense Department traditionally has favored the deployment of submarine cable facilities to complement satellite facilities. In addition, the Defense Department maintains extensive Government-owned facilities to provide international

44 *See generally* 1984 U.S. INDUSTRIAL OUTLOOK at pp. 23 et seq.

45 *See* Statement of Lieutenant General W.J. Hilsman, Director, Defense Communications Agency, Before the Senate Communications Subcommittee in Hearings on S. 2469, 97th Cong., 2d Sess. 96 (1982).

communications. Furthermore, current national security telecommunications policy assigns priority to the creation of a survivable telecommunications infrastructure to support the Federal Government's critical domestic and international telecommunications needs. Additional international satellite facilities would contribute to the »mix of media« national defense requires. Under the limitations proposed here, it is unlikely there would be any significant adverse effects on INTELSAT or other international communications facilities. Accordingly, overall national security telecommunications capability would benefit.

The Defense Department also has a strong interest in the continued strength and vitality of the U.S. satellite communications and aerospace industries. The Defense Department has expressed concern that the United States not become dependent on foreign-owned or controlled firms to provide necessary services and equipment.[46] Approval of the satellite system applications now pending before the FCC, subject to limitations, would advance U.S. technology, and defense interests would benefit.

The Defense Department could benefit significantly from changes in FCC rules to facilitate cost-based access to INTELSAT. Access by firms other than Comsat has been authorized by the FCC in the past.[47] Such access is a means by which international communications costs can be substantially reduced and service flexibility improved. Both would benefit the Defense Department as a major user.

In conclusion, the Defense Department and the other parts of the national defense and security community have a strong interest in the future economic strength and technological vitality of the INTELSAT system. Under the limited entry approach discussed here, those legitimate interests would be protected. Indeed, authorizing additional U.S.-owned and controlled international satellite systems could further defense interests by improving the survivability of the U.S. national telecommunications infrastructure and maintaining an effective and efficient aerospace industry.

Limitations on International Service Are Sustainable

The President's national interest determination stated that certain criteria were necessary to ensure that the United States meets its international obligations and to further its telecommunications and foreign policy interests. The Secretaries of State and Commerce have informed the FCC that, in addition to INTELSAT consultation, final authorization of each system must restrict such licensee to providing services through the sale or long-term lease of transponders or space segment capacity for communications not interconnected with public-switched message networks (except for emergency restoration service).

While recognizing the public benefit of these restrictions, some industry participants have expressed concern that the FCC may not have the power to impose or maintain such limits. They point particularly to the reversal of the FCC in the so-called *Execunet* decision.[48] Their concern is misplaced, however, for if based on proper regulatory

46 *Id.* at 93–94.

47 *See* Transiting Decision, 23 F.C.C.2d 9, 30 F.C.C.2d 513 (1981); *see also* ITT World Communications, Inc. v. FCC, 725 F.2d 732, 752 n. 48 (D.C. Cir. 1984).

48 MCI Telecom. Corp. v. FCC, 561 F.2d 365, 580 F.2d 590 (D.C. Cir. 1977, 1978). *See generally* Hutton, *The Proposed Deregulation of Domestic Common Carrier Telecommunications*, 69 CAL. L. REV. 455, 457 (1981); Warren, *Intercity Telecommunications Competition After Execunet*, 31 FED. COM. B.J. 117, 129 (1978).

procedures and findings, FCC limitations on international service offerings by new satellite entrants are sustainable.

Applicants to construct and operate satellite systems are subject to title III of the Communications Act as previously noted, and many of the provisions of that title broadly empower the FCC to take the actions required here. Section 301 prohibits persons from transmitting radio signals except in accordance with the Act and with a license granted under its provisions. Section 303(b) authorizes the Commission to prescribe the nature of the service to be rendered by each class of licensed station and each station within a class.

Under section 303(f), the FCC is authorized to adopt regulations necessary to carry out the provisions of the Act. Section 303(r) specifically authorizes the FCC to prescribe such restrictions and conditions as may be necessary to carry out the Act or U.S. obligations under treaties or conventions relating to radio or wire communications.

In addition, section 308(c) provides that in granting a radio license for commercial communication between the United States and any foreign country, the FCC may impose any terms, conditions, or restrictions authorized to be imposed under section 2 of the Submarine Cable Landing Act (47 U.S.C. 35). Again this empowers the Commission to withhold, revoke, or condition a license.[49]

Section 309(h) states that each license in subject to conditions, including that the licensee does not have a right to operate the station beyond the term of the license nor in any manner other than authorized therein.

With this extensive statutory support,[50] the courts naturally have found a delegation of wide discretion to the Commission: »[I]t is clear that Congress meant to confer ›broad authority‹ on the Commission ... so as ›to maintain, through appropriate administrative control, a grip on the dynamic aspects of radio transmission‹«.[51]

Despite such underpinnings, some maintain that the FCC's ability to circumscribe the range of services offered by additional international satellite systems is limited, based on their reading of the *Execunet* rulings. In 1976, MCI began marketing a long-distance service called »Execunet.« The FCC determined that Execunet was »message telephone« service (MTS) not »private line« service, that MCI had been limited to providing only specialized or private line services, and thus ordered the offering discontinued. The basis of the FCC's opinion was that there was an implied restriction in the license limiting MCI to specialized services, because the Commission had a written policy of prohibiting specialized carriers from providing MTS service. The court remanded the FCC's decision, because it had not made a specific determination in granting MCI's license that the public interest and necessity required such a restriction. In arriving at its ruling, the court discussed the authority of the FCC to restrict licenses:

49 Functions vested in the President by section 35 of the Cable Landing Act were delegated to the FCC by Executive Order 10530, 3 CFR 189 (1954–1958 Comp.).

50 Similar authority has been granted to the FCC under title II of the Act with respect to common carriage. For example, »[t]he Commission shall have the power to issue such certificate as applied for, to refuse to issue it, or to issue it ... for the partial exercise only of such right or privilege, and may attach to the issuance of the certificate such terms or conditions as in its judgment the public convenience and necessity may require.« 47 U.S.C. 214(c).

51 FCC v. Pottsville Broadcasting Co., 309 U.S. 134, 138 (1940), quoted in FCC v. Midwest Video Corp., 440 U.S. 689, 696 (1978).

> [T]he usual way in which a carrier becomes restricted in the services it may offer is for the Commission to write restrictions into the facilities authorizations that must be obtained pursuant to Section 214 of the Communications Act before any communications line may be built, operated, or extended. Accordingly, a carrier can usually tell if it is subject to service restrictions simply by examining the instruments of authorization issued to it by the Commission.[52]

The court did not find the FCC lacked authority to prescribe the services MCI could offer, but only that when granting MCI its authorization the FCC had not followed proper procedures and made the requisite public interest finding that such service limits were appropriate. Assuming the FCC were to make proper findings in the case of each of the proposed new international satellite systems. new entrants can legally be circumscribed in the range of services they may offer.[53] This is especially true since the President has determined that such limitations are required for foreign policy and related reasons, an area in which the courts have generally deferred.[54]

Opponents of the pending applications argue any limitations placed on new entrants ultimately might be relaxed domestically. Changed circumstances conceivably might lead to such reconsideration in the future; U.S. domestic common carrier regulations in general have tended to be liberalized over time. The same is not true abroad, however. Virtually all European PTTs currently enforce service restrictions, and there are few indications this will change. Enforcement measures include on-site monitoring of users' telecommunications centers, and use of facilities for unauthorized purposes is grounds for discontinuation of service. Most European PTTs also do not permit use of customer-premises earth stations at this time, nor the resale of communications circuits. U.S. international firms also often admonish their customers not to use facilities for impermissible services.[55]

Given the multilateral nature of international telecommunications and the fact customers of the proposed new U.S. satellite systems will be obliged generally to deal through local PTTs for the foreseeable future, we believe limitations on the services offered by

52 561 F.2d at 373.

53 The FCC has successfully exercised similar authority a number of times, for example restricting the scope of AT&T and Comsat's participation in domestic satellite services (Domestic Communications-Satellite Facilities, 35 F.C.C.2d 844, 853 (1972)) and restricting the Satellite Business Systems (SBS) joint venture of IBM and Comsat (Satellite Business Systems, 62 F.C.C.2d 997, 1046, *recon. denied*, 64 F.C.C.2d 872, 873 (1977)).

54 *See, e.g.*, Dames & Moore v. Regan, 453 U.S. 654, 678 (1981); Haig v. Agee, 453 U.S. 280 (1981).

55 One of the leading providers of international data processing services informs customers of its sophisticated »Cybernet Services,« for example –

> Users of Control Data services should be aware that the rules and regulations of the United States and International Telecommunications Regulatory Agencies prohibit Control Data from using communications services it leases from domestic, international and foreign communications carriers to transmit information for its users which is not part of a »single integrated« data processing service. All information transmitted must be directly related to the data processing applications or service provided by Control Data and unprocessed information shall not be allowed through the service between user terminals, either directly or on a store and forward basis. Noncompliance with these rules and regulations may force Control Data to discontinue the users' data processing service.

Cybernet Reference Manual (cover sheet) (1980, rev.).

new systems can be effective. If there were sufficient noncompliance with the FCC's restrictions to raise the prospect of significant economic harm to INTELSAT, such noncompliance would almost certainly be obvious to competitors and regulators alike. As indicated, no regulatory regime whether here or abroad can ever achieve 100 percent effectiveness nor be immune to further evolution. The limitations proposed here, however, will prove sufficiently effective to prevent any significant adverse impact on INTELSAT. If changes in the U.S. limitations are undertaken in the future, moreover, those will be accomplished consistent with our INTELSAT obligations.

»Predatory Pricing« and Related Concerns

Some have expressed concerns over possible pricing responses to competition by INTELSAT. Price competition, however, benefits consumers. Price reductions by an established firm with market power are not always or even usually »predatory«, much less socially or economically undesirable. Too rigid or unbending a pricing standard may discourage price cutting, maintain prices in a market significantly above competitive levels, and also induce entry by less efficient firms. Too flexible a standard obviously could permit a firm with substantial market power to reduce price below actual cost and thus damage or inhibit competition. Nevertheless, we believe that concerns about possible predatory pricing are premature. The economic and legal literature provides very little evidence predatory pricing has ever occurred.[56]

INTELSAT's ability to engage in predatory pricing, in any event, is dependent in large part on the willingness of the U.S. Government to overlook such conduct or to fail to take remedial steps if it occurs. Any such assumption, however, is obviously flawed. If it were shown, for example, that INTELSAT was charging rates for customized offerings which it could not cost-justify and which were significantly injuring U.S. competitors, the Government would necessarily reexamine the restrictions placed on U.S. entrants pursuant to the President's national interest determination and take appropriate remedial actions.

VI. *Conclusion*

The applications to establish additional international satellite systems now pending before the FCC presented four options. The Executive could have recommended (1) approval, (2) denial of the applications outright, (3) approval of the applications subject to specific qualifications, or (4) further study, with postponement of any decision for an indefinite period. The unanimous view among the member agencies represented on the SIG is that it would be in the U.S. national interest to allow new providers of international satellite facilities, provided INTELSAT were not exposed to significant economic harm. The President's determination reflects this view.

56 *See, e.g.*, McGee, *Predatory Price Cutting: The Standard Oil Company Case*, 1 J. LAW & ECON. 137 (1958); Telser, *Cut-Throat Competition and the Long Purse*, 9 J. LAW & ECON. 259, 267 (1966). *See also* Berkey Photo, Inc. v. Eastman Kodak Co., 603 F.2d 262, 273, 294 (2d Cir. 1979); Northeastern Teleph. Co. v. AT&T, 651 F.2d 76, 93 (2d Cir. 1981).

There is sufficient risk of significant adverse economic impact on INTELSAT to make blanket approval of unrestricted competition unwise. It would also be premature to take such a step until the results of cost-based access, new fiber optic cables, and new INTELSAT services are fully evaluated. Unrestricted entry could ultimately undermine the economic integrity of this important international enterprise, which would be inconsistent with the U.S. national interest.

The case has not been made for flatly disapproving the existing applications. The new entrants have made a threshold showing that services they propose are not now available on comparable terms. Limited entry along the lines recommended would further U.S. international trade interests, promote technological progress, and be consistent with national defense and security interests as well. Given these limitations, and the restrictions likely to be placed on any new satellite system by telecommunications authorities abroad, the risk of any significant adverse impact on INTELSAT is exceedingly small.

Further study and resulting delay is unlikely to further the national interest. Over a year of extensive study and review by the Executive branch has already taken place. This review has not resulted in the submission of credible information supplied by anyone, including INTELSAT and Comsat, which demonstrates plausible adverse effects. There is no basis to assume such information will be forthcoming.

Satellite systems entail significant lead time. Time is required to secure the requisite spacecraft, to reach launch agreements, and to secure operating arrangements. U.S. regulatory procedures are generally more time consuming than those abroad, where decisions can sometimes be reached and implemented without the regulatory proceedings and protracted court appeals characteristic of U.S. regulation. Consultation with INTELSAT is also required. Even were the pending applications approved by the FCC immediately, service would not be available for some time.

Government should not stifle private entrepreneurial initiatives absent sound and compelling public policy reasons. Such initiatives should not be discouraged when the services proposed could prove of value to customers, improve their productivity and efficiency, and thus enable American firms to compete more effectively both at home and abroad. The public policy case for continuing the status quo and flatly prohibiting additional international satellite systems is weak. Simply the pendency of U.S. applications has caused INTELSAT to accelerate plans for special business-oriented services and has precipitated a beneficial review of competitive conditions in the international satellite field generally. Further study and inevitable delay are unlikely to yield public dividends commensurate with the economic costs imposed.

It is the view of the Executive branch that the national interest will be furthered by approving additional international communications satellite systems subject to limitations designed to minimize adverse effects on INTELSAT. Specifically, additional systems should be restricted to providing services through the sale or long-term lease of transponders or space segment capacity for communications not interconnected with public-switched message networks (except for emergency restoration service). Consultation must be undertaken with INTELSAT pursuant to Article XIV(d) of the Definitive Agreement.

Appendix A

The White House
Washington

November 28, 1984

Presidential Determination
No. 85–2

Memorandum for The Secretary of State
The Secretary of Commerce

By virtue of the authority vested in me by the Constitution and statutes of the United States, including Sections 102(d) and 201(a) of the Communications Satellite Act of 1962, as amended (47 U.S.C. 701(d), 721(a)), I hereby determine that separate international communications satellite systems are required in the national interest. The United States, in order to meet its obligations under the Agreement Establishing the International Telecommunications Satellite Organization (INTELSAT) (TIAS 7532), shall consult with INTELSAT regarding such separate systems as are authorized by the Federal Communications Commission. You are directed jointly to inform the Federal Communications Commission of criteria necessary to ensure the United States meets its international obligations and to further its telecommunications and foreign policy interests.

This determination shall be published in the *Federal Register*.

Appendix B

Honorable Mark S. Fowler
Chairman
Federal Communications Commission
Washington, D.C. 20554

November 28, 1984

Dear Mr. Chairman:

The President has determined that separate international communications satellite systems are required in the national interest. He has also directed that we inform the Federal Communications Commission of criteria necessary to ensure the United States meets its international obligations and to further its telecommunications and foreign policy interests. Prior to final authorization by the Commission of any systems, to assure that the United States meets its obligations as a Party to the Agreement Establishing the International Telecommunications Satellite Organization (INTELSAT) (TIAS 7532):

(1) each system is to be restricted to providing services through the sale or long-term lease of transponders or space segment capacity for communications not interconnected with public-switched message networks (except for emergency restoration service); and,

(2) one or more foreign authorities are to authorize use of each system and enter into consultation procedures with the United States Party under Article XIV(d) of the INTELSAT Agreement to ensure technical compatibility and to avoid significant economic harm.

The President's determination, its conditions, and these criteria are premised on our review of the issues prompted by the applications now before the Commission. If proposals substantially different are forthcoming, further Executive Branch review may be required.

The Commission should afford interested parties an opportunity to submit timely comments on the pending applications in view of these Executive Branch recommendations.

A memorandum of law concerning Article XIV of the INTELSAT Agreement is enclosed.

Sincerely,

Secretary of State | Secretary of Commerce